Jürgen Valldorf · Wolfgang Gessner (Eds.)

Advanced Microsystems for Automotive Applications 2007

Jürgen Valldorf · Wolfgang Gessner (Eds.)

Advanced Microsystems for Automotive Applications 2007

With 196 Figures and 17 Tables

Springer

Dr. Jürgen Valldorf
VDI/VDE Innovation + Technik GmbH
Steinplatz 1
10623 Berlin, Germany
valldorf@vdivde-it.de

Dr. Wolfgang Gessner
VDI/VDE Innovation + Technik GmbH
Steinplatz 1
10623 Berlin, Germany
gessner@vdivde-it.de

ISBN-13 978-3-642-09043-1 e-ISBN-13 978-3-540-71325-8

Springer is a part of Springer Science+Business Media

springer.com

Cover design: WMXDesign, Heidelberg

Printed on acid-free paper 68/3100/YL - 5 4 3 2 1 0

Preface

European automotive industry is currently facing a series of challenges due to changing determinants of world markets and of a globalized organization of added value production. Looming problems of primary energy and materials availability as well as being subject to significant societal demands, particularly in terms of road safety and environment characterise the situation of car production today.

The current debate on global warming is accelerating the automotive industry's efforts of reducing emissions, particularly CO_2. Already today progress is visible: mid sized cars equipped with 105 kW diesel engines reach CO_2 values of little more than 120 g/km. Fast reactions and the enormous R&D investments place the automotive industry among the most innovative industries in facing greenhouse gas emissions. Visions of zero emission cars will soon become a reality by the introduction of disruptive technologies such as: the „in-wheel motors electrical vehicles" and super-batteries combining the properties of both batteries and super-capacitors. Smart miniaturized systems are the key for technological breakthroughs at systems level.

Road safety is a permanent issue of highest priority in Europe as the human and economic costs of road fatalities and injuries remain significant. Zero fatalities is the long term vision while retaining the liberty of individual driving. Passive and active systems include information on vehicle dynamics limitations and intervention if necessary, personalized safety systems adapted to the characteristics of the individual, driver drowsiness monitoring, collision avoidance and mitigation systems, etc. The objective is to provide optimal driver support by the technical systems taking into account of vehicle and driver capabilities and characteristics.

Microsystems in this context become advanced – as it is expressed by the title of this publication. They become smart and complex systems constitute the core of the mentioned new features. The 2007 publication on Advanced Microsystems for Automotive Applications highlights a series of aspects and newest developments facing economic and societal requirements. The aim is also to create awareness for new developments in which context new smart technologies are co-determining radical societal changes and for the need of new innovative forms of policy-industry cooperation.

The publication in hand is a mirror of the International Forum on Advanced Microsystems for Automotive Applications which continues to be a unique exchange forum for companies in the automotive value chain. The 11th event of a series starting in 1995 took place in Berlin on 09./10.05.2007.

I like to express my sincere thanks to the authors for their valuable contributions to this publication. I like to thank the members of the AMAA Honorary Committee and the Steering Committee for their advice and continuous assistance. Particular thanks are addressed to the European Commission for their financial support through the Innovation Relay Centre Northern Germany and to the supporting companies Robert Bosch GmbH, First Sensor Technology, IBEO and VTI Technology. Tribute has also to be paid to EPoSS, the European Technology Platform on Smart Systems Integration.

Last but not least, my thanks are addressed to the Innovation Relay Centre team at VDI/VDE-IT for its strenuous engagement, to Mr. Michael Strietzel for preparing this book for publication, and particularly to Dr. Jürgen Valldorf, project manager and chairman of the AMAA.

Berlin, May 2007

Wolfgang Gessner

Public Financers

Berlin Senate for Economics and Technology

European Commision

Ministry for Economics Brandenburg

Supporting Organisations

Investitionsbank Berlin (IBB)

mst*news*

ZVEI - Zentralverband Elektrotechnik- und Elektronikindustrie e.V.

Hanser automotive electronic systems

Micronews - The Yole Developpement Newsletter

enablingMNT

Co-Organisators

European Council for Automotive R&D (EUCAR)

European Association of Automotive Suppliers (CLEPA)

Advanced driver assistance systems in Europe (ADASE)

Honorary Commitee

Eugenio Razelli	President and CEO Magneti Marelli S.P.A., Italy
Günter Hertel	Vice President Research and Technology DaimlerChrysler AG, Germany
Rémi Kaiser	Director Technology and Quality Delphi Automotive Systems Europe, France
Nevio di Giusto	President and CEO FIAT, Italy
Karl-Thomas Neumann	CEO, Member of the Executive Board Continental Automotive Systems, Germany

Steering Commitee

Dr. Giancarlo Alessandretti	Centro Ricerche FIAT, Orbassano, Italy
Mike Babala	TRW Automotive, Livonia MI, USA
Serge Boverie	Siemens VDO Automotive, Toulouse, France
Geoff Callow	Technical & Engineering Consulting, London, UK
Bernhard Fuchsbauer	Audi AG, Ingolstadt, Germany
Kay Fürstenberg	IBEO GmbH, Hamburg, Germany
Wolfgang Gessner	VDI/VDE-IT, Berlin, Germany
Roger Grace	Roger Grace Associates, Naples FL, USA
Dr. Klaus Gresser	BMW Forschung und Technik GmbH, Munich, Germany
Henrik Jakobsen	SensoNor A.S., Horten, Norway
Horst Kornemann	Continental Automotive Systems, Frankfurt am Main, Germany
Hannu Laatikainen	VTI Technologies Oy, Vantaa, Finland
Dr. Torsten Mehlhorn	Investitionsbank Berlin, Berlin, Germany
Dr. Roland Müller-Fiedler	Robert Bosch GmbH, Stuttgart, Germany
Paul Mulvanny	QinetiQ Ltd., Farnborough, UK
Dr. Andy Noble	Ricardo Consulting Engineers Ltd., Shoreham-by-Sea, UK
Dr. Ulf Palmquist	EUCAR, Brussels, Belgium
David B. Rich	Delphi Delco Electronics Systems, Kokomo IN, USA
Dr. Detlef E. Ricken	Delphi Delco Electronics Europe GmbH, Rüsselsheim, Germany
Jean-Paul Rouet	Sagem SA, Cergy Pontoise, France
Christian Rousseau	Renault SA, Guyancourt, France
Patric Salomon	4M2C, Berlin, Germany
Ernst Schmidt	BMW AG, Munich, Germany
John P. Schuster	Continental Automotive Systems, Deer Park IL, USA
Dr. Florian Solzbacher	University of Utah, Salt Lake City UT, USA
Bob Sulouff	Analog Devices Inc., Cambridge MA, USA
Berthold Ulmer	DaimlerChrysler AG, Brussels, Belgium
Egon Vetter	Ceramet Technologies Ltd., Melbourne, Australia
Hans-Christian von der Wense	Freescale GmbH, Munich, Germany

Conference chair:

Dr. Jürgen Valldorf	VDI/VDE-IT, Berlin, Germany

Table of Contents

Market

Safety

Components and Generic Sensor Technologies

Appendices

Market

Pressure Monitoring Sensing Market - Who will win the Race between Hybrid vs Monolithic Integrated Products?

M. Potin, J. Ch. Eloy, E. Mounier, Yole Developpement

Abstract

Automotive applications are still driving new MEMS development for new applications. Pressure monitoring application is one of the very dynamic one in particular. This article highlights the evolution of the pressure monitoring sensors both at the component and module level. New product announcements have recently fragmented the product offer in two kinds: Hybrid modules vs. single die packaged modules. This change will strongly affect current player position. This paper will present market needs, status, and trends, to discuss which players will win the race.

Starting with an historical reminder of TPM application, current OEM needs will be described followed by 2006 - 2010 market estimation. In particular, we will focus the discussion on the evolution of the s functions: which level of integration will s support. The final part of the presentation will highlight how the new TPM product offering will impact the industrial chain.

1 Historical Start of TPMS

The measurement of the pressure of the using MEMS based pressure sensors (pressure monitoring systems or TPMS) has started as a real application in 1985: the application was at that time mainly linked to trucks and heavy vehicles but also for high end vehicles (Porsche was the first to implement such system in the 959 model). For example, most of the used in trucks are not acquired by the company using the truck but are rented for a specific duration (for example 1 Million kilometres): the companies providing the service of truck s renting (a subsidiary of Michelin or Firestone for example) had implemented TPMS in order to be able to change s before problems appears. It was during 10 years roughly a market of million units per year.

The market has really started after the Firestone/Ford series of problems 5 years ago: more than 100 lethal accidents have taken place in the USA, pushing the US regulatory body (the NHTSA, National Highway Traffic Safety Administration) to define a rule (Rule No. FMVSS 138). After 3 years of discussions and modifications, the almost final regulation has been edited beginning of 2005. The regulation is not defining the system to be used (direct measurement or indirect measurement like the re-use of the ABS systems) but the performances to be obtained are clearly indicating that the direct measurement is the solution to be adopted.

2 Why the Need for Pressure?

While the main driver of TPM systems implementation is regulation in US, Europe and Asia is sensitive to the benefits in safety and cost savings enabled.

- **Safety:** Pressure monitoring helps maintaining proper tire inflation reducing blowouts, mitigating hydroplaning, reducing braking distance and car handling.
- **Cost Savings:** Proper tire inflation shows increased life and reduces gas mileage. Savings are even better identified by professional applications (trailers, tractors, ...) where tires are a major cost for commercial vehicles.

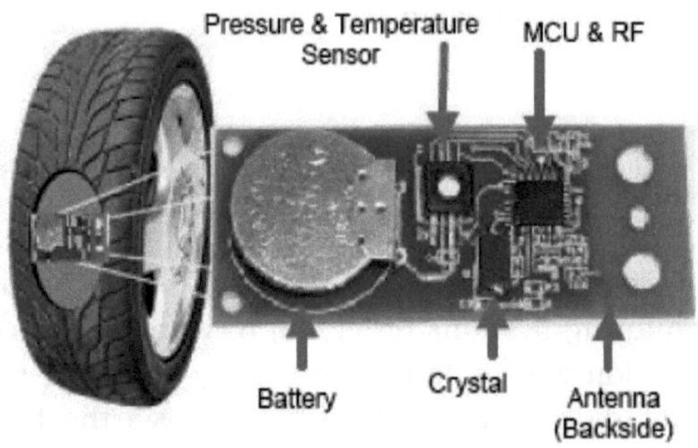

Fig. 1. TPM system having showing separate pressure sensor and processing and transmitting electronics (courtesy: Motorola)

It is still a question for major tire equipment manufacturer to understand if an EU regulation would happen. It is likely that similarly with safety features like stability control, the European automotive market will progressively implement this technology without such incentive.

3 TPM System Description

The TPM systems include the following devices: pressure sensing, signal processing, RF transmission, battery, and an antenna. The easiest location to place this sensor is the valve, where the systems can be plug on it or build with it.

The systems monitor the pressure within the four vehicle tires and transmit it to the receptor located at the central area of the car.

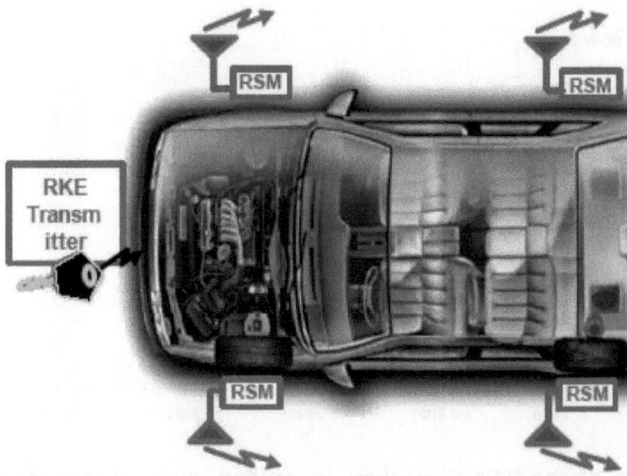

Fig. 2. Location of the TPMS modules on car (courtesy: Motorola)

OEMs want these systems to easily adapt both to the actual tire structure while not modifying their assembly lines. Its localization at the tire valve is today available within two forms to be plugged or screwed on the wheel. Most OEMs are today asking for devices to be plugged because it reduces assembly time.

Most of the systems are built in equipment today in cars. First after sales TPM kits are available since this year from Schraeder. This is rather consider as a niche market because not ruled by the THREAD act.

4 Current Market Needs

The content of the legislation is clear: by September 2007, 100% of the vehicles sold on the US markets will have to have TPMS systems. That means a market of 60 million units in 2008 for the US market only. The legislation is also defining a minimum of 20% implementation in 2006 with a carry back/carry forward systems.

So the US market is clearly driven by the legislation. Outside USA, the market is driven by the adoption by the customer: the TPMS systems in Europe (and step-by-step in Japan) is implemented in high end cars now and is migrating to medium to low end cars step by step.

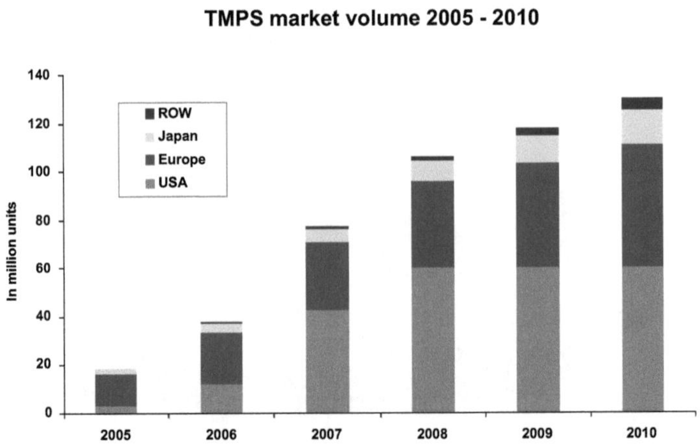

Fig. 3. Evolution of the TPMS markets in million silicon based devices (source: Yole Développement)

We can expect that the markets outside USA will be in the range of 45 million units in 2008. The (Fig. 3) is providing our analysis of the TPMS market evolution outside USA (data extracted from our report "Status of the MEMS industry 06"). Due to the fact that no regulation is indicating which measurement solution to be chosen, it is likely to be a mix between direct and indirect measurement.

The main trends is today towards the direct measurement but several companies already provide or are in development of indirect systems for a mid term introduction. The current solution based on wheel speed differential measurement (named Deflation Detection System DDS at Continental) is considered as a low cost alternative to direct measurements. Complex indirect measurement systems, with superior performance providing car body dynamic data too are at development stage at several locations.

Among all pressure sensors applications, TPMS will have the largest market growth over 2005 - 2010. The market CAGR will be of 36%. Among several market drivers, the trends for a future regulation should be investigated.

5 Main Players

Several equipment manufacturers have tried to enter this market and the companies which are successful are the keyless entry system manufacturers: they are re-using the wireless structure for the TPMS. All the companies involved on this market (Schräder, Beru, ...) are coming from this field.

Several players are providing the MEMS sensor:

- **Infineon/Sensonor**: Sensonor is by far the world leader. The company has been acquired by Infineon in order to have access rapidly to this product middle of 2004. Sensonor is producing more than 2,4 million units per month (in beginning 2006). The production level has already jumped four times above the first month of 2005 level. The key strength of Infineon/Sensonor is to have been able to propose at the beginning of this market a module, including the sensor, the battery and the wireless module. Using this module, the keyless entry systems manufacturers have been able to enter this market more rapidly. We estimate at Yole Développement that Sensonor has more than 60% of the world market. In each module, a simple accelerometer is also included in order to be able to detect the movement of the tire start the measurement.
- **GE Novasensor**: Novasensor is number two on this market, mainly working with Schräder.
- several other companies are proposing devices like SMI (USA) or are developing devices (Bosch, Siemens VDO, Freescale, ...).

6 Major Change: New Product Introduction

While current players are today very involved in new product generation development, new comers are even more thinking at extending TPM system features to be at the heart of the car monitoring system: the "intelligent " concept.

6.1 Short Term Product Evolution

New product announcement has recently fragmented the product offer in two kinds: Hybrid modules vs. single die packaged modules.

New product announcements were linked with the 3rd generation products: the full monolithic approach. Pressure sensor, ASIC, microcontroller, RF, and sleep mode sensor (accelerometer) are integrated on a single package. Several companies have shown such devices at in Detroit, at Convergence in October 2006, like Infineon and Freescale. Development roadmap of Motorola see below is still valid. This integration approach will help reducing component cost while saving total module weight to stay under the 40 g barrier with additional functions.

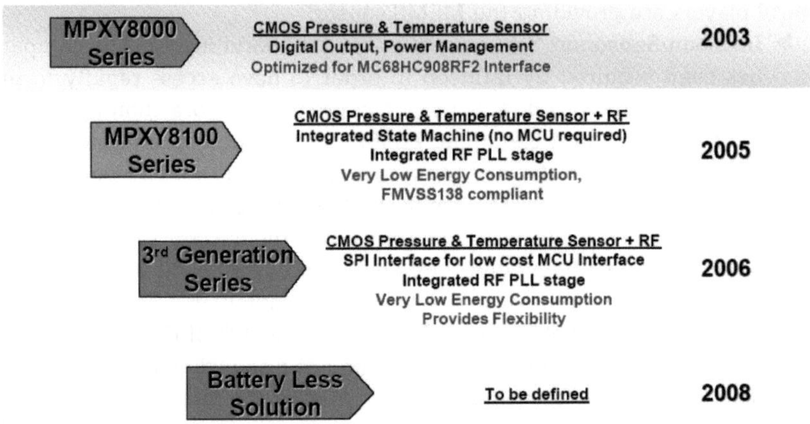

Fig. 4. Motorola 2003 TPMS roadmap

Hybrid devices are still very popular to some manufacturers like Schrader, TI, SMI, Motorola Automotive (now part of Continental), Hitachi. These players are convinced a hybrid system would bring the flexibility to optimize each component to reach better low power performances. But no test results are today public!

6.2 Long Term Vision

What is extremely interesting is that different players like Infineon, Michelin and several companies in Japan are developing several concepts in order to change the role of the tire. According to their vision, the tire will be more and more "Intelligent" and the centre for several measurements. The Fig. 2 is presenting this concept. The tire will integrate more sensing capabilities:

- inertial sensing in order to be able to detect early drift of the car and interact with electronic stability systems
- speed of the wheel
- traction
- pressure and temperature

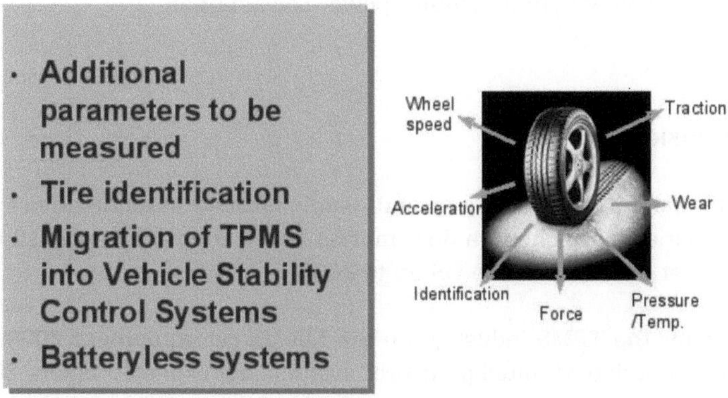

Fig. 5. On the way to an "intelligent tire" (source: Infineon)

No products are today available but the major manufacturers and TPMS sensor manufacturers are working in order to be able to implement more and more functions for improved safety.

We just hope that it will still be simple to change a wheel when a problem occurs: being obliged to have specialised technicians in order to be able to change a wheel can clearly stop the adoption of such innovation. But the tire will certainly be the centre of major evolutions in the future. The TPMS application is one of the fastest growing applications of MEMS devices and certainly a very interesting business case.

7 Who will the Winners?

The key success factors of OEMs to select TPM systems are

- cost of the module
- system life time without battery replacement
- self organizing system with no calibration

This forecasted market dynamics has recently changed the competitive landscape. The original niche players and products today face the 2008 future product announcement of large IC players.

Our analysis, at Yole Developpement, shows that players are following two development strategies. The goal of the large IC players is clearly to take a large share of this market to absorb the huge development effort realized. We believe the success will strongly depend from the ability of the single package modules to go below hybrid modules power consumption.

8 Conclusion

The TPMS market is one of the most dynamic markets in automotive applications. It is expected to shows a 36% market CAGR from 2005 to 2010 with a strong market evolution in the US up to 2008.

Next steps for the TPMS industry are the full US car equipment 2008 status and the introduction of "intelligent tire" systems not expected before 2010.

Mathieu Potin Jean-Christophe Eloy, Eric Mounier
Yole Développement
45 rue Sainte Genevieve
69006 LYON
FRANCE
potin@yole.fr

Keywords: market, application, automotive, TPMS, tire, pressure, sensor

Prospects and Strategic Considerations for Automotive MEMS Component Suppliers and System Integrators

R. Dixon, J. Bouchaud, WTC – Wicht Technologie Consulting

Abstract

What are the opportunities for MEMS sensors components and what are the new market drivers. What are the opportunities for and current challenges facing MEMS suppliers, and what does this mean for the proportion of income invested in R&D? How does this affect the industry chain and how can suppliers stay on top? To answer these questions we look at market trends in the automotive sector and examine the changing level of involvement of systems and component manufacturers in MEMS along the automotive supply chain.

1 Overview of Automotive MEMS Sensors Applications

A large number of sensors are integrated in the modern car (Fig. 1). Among these are:

- pressure sensors for engine management and air intake monitoring, e.g. manifold and barometric air pressure measurements, tire pressure monitoring systems (TPMS) or as side airbag sensors
- accelerometers for air bags and combined for example with gyroscopes for Electronic Stability Program or ESP, GPS dead-reckoning navigation and roll detection applications
- inclinometers for zero set for gyro, electronic parking brake
- flow sensors for HVAC applications (under investigation)
- infrared sensors for driver vision enhancement, cabin temperature, anti-fog

The market for MEMS sensors was $1.53 billion in 2006 and we forecast this to exceed $2 billion in 2010, a compound average growth rate of 9% in dollar terms. The main applications today, in order, are pressure sensors, precision gyroscopes, and low- and high-g accelerometers. This order will be preserved in future, with the possible exception of flow sensors, which we are currently investigating. The total number of sensor units will grow from over 370 million to well over 520 million sensors (CAGR = 12%) with price erosion per piece running as high as 4-5% per year depending on the component.

The main application areas are airbags, vehicle dynamics (ESP) and TPMS. The airbag market is slowing but still has life as a result of opportunities for satellite bags. More sophisticated airbag solutions are also being developed as a result of passenger occupation detection and combined pressure/accelerometer side airbags.

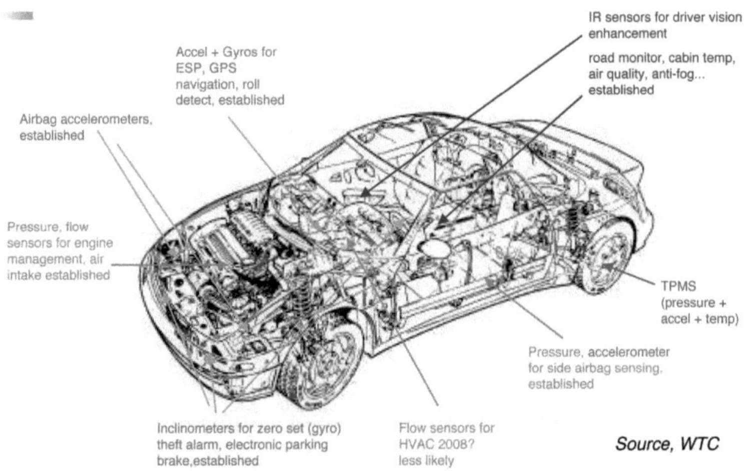

Fig. 1. Overview of MEMS applications in automotive applications (Source: WTC)

There are new opportunities that will accelerate the market beyond 2010. We are currently updating our market data to take account of issues such as the mandatory introduction of ESP vehicle dynamic systems in the USA in 2010. This event alone will rapidly accelerate the market for such systems, producing an interesting battleground for manufacturers as prices are driven down (although ESP systems are still feature relatively costly and challenging to manufacturer sensors). The market volume will therefore not grow correspondingly.

Another example is TPMS, which will undergo more advanced iterations that eventually target battery-less solutions, migrating first from "in-wheel" through "in-tire" systems – potentially a 1.2 billion unit retro-fit market. Such systems will benefit from additional sensors as they add new functionalities such as force, friction/grip, tire ID and wheel speed to feed more intelligent ESP systems and other types of sensor fusion scenarios.

2 Opportunities and Challenges for MEMS Component Suppliers

There are opportunities for MEMS component suppliers to capitalize on new norms such as the mandatory introduction of ESP systems in the US from 2010. This is an opportunity but can also be perceived as a market damper as it transitions from high value-added to commodity product, and those that manufacture these complex systems lose a level of product differentiation. Nevertheless it opens a great market opportunity for those who can anticipate the new regulations, and increasing the R&D level in anticipation is a good strategy.

In addition, we expect opportunities from new products and advanced iterations, a case in point being TPMS. The latter is a possible driver for the adoption of energy scavenging technology in the automotive sector. Meanwhile, new players in the market would also provide challenges, although we believe that only the big players like STMicroelectronics will have a serious chance to break into this sector in contrast to smaller start-ups.

Clustering or sensor fusion will also not likely dramatically reduce the number of MEMS sensors. Signal fusion is a more likely scenario, and sensors will still be required in different positions of the car to provide sensing information. Ultrasound sensors represent a likely replacement scenario for some accelerometers in front crashes, and there is a limit to the number of airbags, but new applications continually emerge, e.g. electronic parking brakes.

Staying with market brakes for a moment, the MEMS sensor market could see the eventual saturation of passive safety features, with mostly non-MEMS active safety systems coming to the fore. Another market damper for sensors could be the increasing cost of the average car as the number of advanced safety and comfort features continues to rise, with consumers becoming more choosy concerning what features they are prepared to pay for.

Finally, a continuing challenge to component suppliers is the strong price pressure on sensors and other components. This runs at 4-5% annually.

What are implications for the industry chain and how do component manufacturers and systems suppliers change these challenges into opportunities?

3 Implications for Supply Chain

Representative companies involved in the automotive supply chain are shown in Fig. 2. While groups regularly converse on topics like micro-controller units, the general perception that car manufacturers are not involved in MEMS sensor development is no longer the case. Instead of providing just "black box" specifications to sub-system (tier 1) suppliers, these car manufacturers begin to talk directly to the sensor component manufacturers.

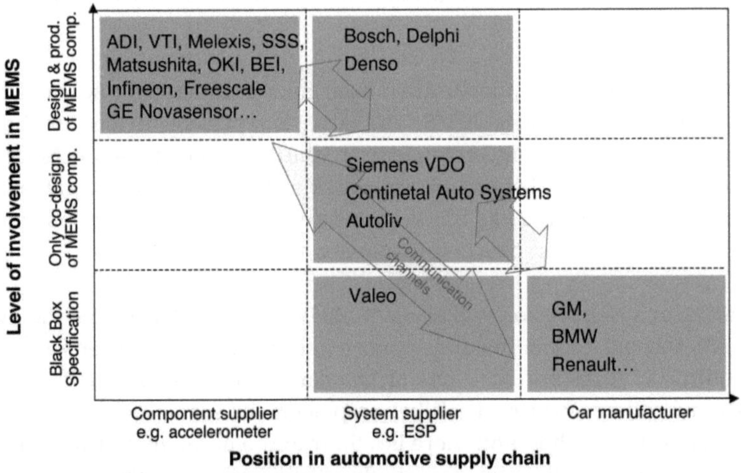

Fig. 2. OEMs are taking increased interest in the activities of the MEMS sensor manufacturers (Source WTC).

There are several fairly obvious reasons for this. First and foremost OEMs are interested in what the technology can do and need to understand it to do so. Secondly, they are taking much greater interest in determining the breakdown for the components among the system suppliers to understand costs better. These tier 1 suppliers of course talk to both groups. Component manufacturers talk to sub-system manufacturers – their direct customers – but also want to understand their customers. The insights from the OEM help component makers to determine their long-term R&D programs and how new generation MEMS sensors will need to function, with which protocol, and within which economic boundary conditions, etc.

Finally, there is also strong interest in sensor "fusion" concepts to reduce sensor redundancy. Most manufacturers believe this to be on the horizon. This requires an understanding of system architecture and sensor positioning, e.g. the BUS system and protocols set by manufacturers.

The level of involvement in MEMS sensors varies with the company. Companies that divested of their component division (GM/Delphi, for example) are now more than ever require the insights these component engineering groups can bring. On the other hand, BMW is an example of a company that sees technology as a unique selling point and has an active team of engineers that follow technology trends closely. The manufacturers of mid-class vehicles are those more likely to be interested in whether a component delivers a certain function, fulfils a particular set of specifications and are less interested in what the technology actually consists of.

Meanwhile, sub-system suppliers of course have their own roles in MEMS. There are different models at these companies:

- Some sub-system companies have developed components for their own sub-systems and now sell these components to others (i.e. Bosch), albeit after a period of exclusivity.
- Some tier 1 suppliers do not own their own fab but nevertheless are involved closely with co-designing MEMS with the MEMS sensor manufacturers, e.g. Siemens VDO with Infineon.
- Systems suppliers with an existing fab do not outsource MEMS production and in some cases continue to invest in these facilities, case in point Bosch. Big semiconductor fabs like TSMC are not expected to play a major role in sensor manufacture. We see only one example of an outsourced fab at Dalsa, which is confined to commodity pressure sensors and not strategic or leading-edge technologies such as accelerometers and gyroscopes. Component manufacturing is still a major source of revenue for component and system suppliers and will remain so in future.

4 Implications for R&D Spending

One of the main challenges is the continue pressure on prices. A strong R&D rate is needed to keep margins, for example:

- At the back end through the use of smart packaging, i.e. smaller packages and even hermetic dies to replace packaging.
- At the front end by working on process stability (yield) in order to save on test.
- Using new front-end processes, e.g. the new porous silicon approach being taken at Bosch, or a shift from bulk to surface micromachining processes to lower costs.
- Moving to larger wafer size, i.e. 200 mm.

At most component and sub-system companies an own MEMS research capability is still seen as a major advantage for technology and market edge, especially for high value-add sensors. The rate of R&D investment among component and sub-system suppliers varies from 10-20% in the automotive industry and reflects the continual need to invest in new equipment, innovate but at the same time maintain existing processes and product lines for 7+ years. For early stage research many companies also work in public funded programs or with academic and research organizations to achieve their aims.

At the lower end of the scale, R&D spending at a level of 10% implies that no major new product diversification is being implemented and that the companies will work purely on gaining margin. In a sector with price erosion running at 20% over a five-year period and long product design cycles, this is a difficult business proposition. Investment in new technology and processes to stay ahead in the automotive sector usually implies an R&D rate of around 12-15%.

Higher R&D spending in the range 17-20% implies new product development and diversification into other fields such as consumer applications, e.g. at Analog Devices and Freescale, and more recently by companies like Bosch and VTI. Accelerometers and pressure sensors lend themselves well to this transition. There are unique challenges to be faced in the CE field, such as very short design and product cycles, fast manufacturing ramp - up and high R&D investments. Nevertheless the rewards are potentially large, and by serving more markets allow excellent economies of scale.

5 Conclusions

The automotive sector will continue to be a major market for MEMS sensors. An own MEMS research capability is still seen as a major advantage for technology and market edge in sensors and the majority of component and system suppliers commonly invest something between 12-17% in R&D activities, reflecting in the level of product development, diversification and / or at the higher level the move on new markets such as consumer. In order to stay ahead there is a healthy level of discourse along the entire sensor and system product chain.

R. Dixon, J. Bouchaud
WTC - Wicht Technologie Consulting
Frauenplatz 5
80331 Munich
Germany
Richard.Dixon@wtc-consult.de
Jeremie.Bouchaud@wtc-consult.de

Keywords: MEMS sensors, research and development, gyroscopes, accelerometers, pressure sensors, ESP, TMPS

Safety

Reduced Stopping Distance by Radar-Vision Fusion

S. Lüke, M. Komar, M. Strauss, Continental Automotive Systems

Abstract

Increased traffic safety depends on the differentiation in warning, steering actuation, braking interventions as well as the possible passive safety measures in critical situations. Continental built up a test vehicle to develop active safety measures based on radar and camera information. The system focuses on rear end collisions and uses next generation automotive CMOS camera and radar technology to avoid collisions or to reduce their impact severity. The paper describes the networking and the benefit of the additional information, generated by sensor fusion in emergency situations. Key aspects are changes in brake preparation and crash adaptation and the influence of the driver's behaviour, compared to conventional beam sensor based safety systems.

1 Introduction

For the improvement of active and passive safety systems it is essential to consider an increased number of obstacles. The goal is to have an increased knowledge about the relevancy and behaviour of these objects, to measure with an increased accuracy and to have a higher level of confidence in this information. Beam sensor based active and passive safety systems can be optimized by fusion with camera systems. The information regarding the environment can be improved depending on level and quality of fusion.

2 Sensors

Today, powerful beam sensors already provide a huge amount of information which can be used for warning, brake or steering interventions and passive safety control measures in emergency situations. This is typically sensed by near or long range radar, lidar sensors or by a combination of these sensors.

To test the proposed algorithms, a high resolution 77 GHz pulse compression automotive radar sensor that scans the azimuth by varying the periodic structure of a leaky wave antenna is mounted in the frontal bumper of a vehicle. The radar sensor detects objects and measures range, range rate (Doppler) and azimuth angle. For objects like cars the lateral and longitudinal extension can be estimated due to the high resolution in azimuth and range. The sensor chooses the most relevant stationary or moving objects depending on their position relative to the predicted path of the host vehicle, their confidence level, and their radar cross-section (RCS). For sensor fusion, an interface with a barely filtered cluster list of radar reflection points is provided.

The chosen CMOS (Complementary Metal Oxide Semiconductor) single camera system is faced forward and mounted inside the host vehicle typically next to the rear view mirror. CMOS technology is driven by the consumer market and offers a high potential of cost reduction compared to CCD (Charge Coupled Device) technology. The camera has a resolution of 752 × 480 pixel and ±20° horizontal field of view. Important for automotive functions is a dynamic range of > 80 db. This allows a good contrast, even if the illumination of different areas in the picture varies. These functions can be e.g. "High Beam Control", "Lane Detection", "Traffic Sign Recognition" as well as "Sensor Fusion".

Tab. 1 shows the data of the camera and the radar sensor and compares the strengths and weaknesses of both systems. The radar sensor provides accurate longitudinal and approximate lateral position and an accurate differential speed of a radar reflecting target while the camera offers fine lateral information.

In addition to this environmental information the information of inertial vehicle sensors i.e. velocity, acceleration, yaw rate, wheel speeds and steering angle can be used.

	camera	*radar*
data base	*pixel measurements*	*cluster list*
measurement area	*±20° up to ~ 50 m*	*±30° up to 60 m and ±9° up to 200 m*
characteristic	*bad distance information*	*good distance information*
	missing relative speed	*good relative speed*
	good angle information	*bad angle information*

Tab. 1. Characteristics of sensor information

3 Combining Radar with Camera

Comfort and safety systems depend on accurate environmental information. Radar-vision fusion provides this information. The combination of radar and camera can be implemented on different pre-processed data levels. A combination on a high data level would be a target validation. Enhanced performance can be reached by fusion on lower data levels. Understanding the image processing as a sensor, a fusion combined with object tracking on low pre-processed data level with an additional object classification algorithm was implemented. The following sections of this chapter show the basic ideas of our fusion and classification concept.

3.1 Data Fusion

The fusion approach handles low level data by using clustered radar data and camera data with pixel positions and width of objects. The fusion is implemented by using a fusion filter consisting of a Kalman and an association filter. With the measurements and the Kalman filter it is possible to estimate the position and dynamics of the objects. This is called object tracking. The Kalman filter requires an object and sensor model. The object model describes the dynamic characteristic of the objects which should be tracked. The sensor model includes the relation of the object states and the sensor measurements, see [3]. The states of multiple objects measured by two sensors are estimated in a central Kalman filter with varying size. The size depends dynamically on the number of sensors and objects. The tracked objects are described in a uniform object description.

Fig. 1 gives an overview of the data flow. The unsorted measurements of the sensors are fed into the association filter. This filter associates the measurements to the corresponding object tracks in the Kalman filter or initializes a new track in the dynamically resized Kalman filter if there is no appropriate track for a measurement. The association is performed by using the predicted measurements and confidence information which are provided by the Kalman filter.

The uniform object description of the objects tracked includes the position (x_W, y_W), speed (v_{xW}, v_{yW}) and width (b_w) of each object. The information of the tracked objects is used for segmentation of the image in relevant and non-relevant areas. The image processing operates in these regions of interest (ROI). To calculate these ROIs, the object positions (x_W, y_W) estimated by the Kalman filter are given to the image processing in a feedback loop. Inside of the ROIs, a Low Level Vehicle Recognition (LLVR) algorithm identifies the object pixel position and width in the image. A simple approach for LLVR is the detection by vertical edges, the shadow below the vehicle and a pixel symmetry check. [1] describes such an algorithm while [9] shows a higher performance LLVR algorithm based on 3D approach. At night it is feasible to detect the lights. The image processing algorithm delivers a list containing pixel position and width of objects in the image. This is fed into the dynamic Kalman filter as an additional input. The fusion of radar and vision based vehicle detection is presented in [10]. To ensure that the measurement is valid and that the target is relevant all tracked, objects are classified.

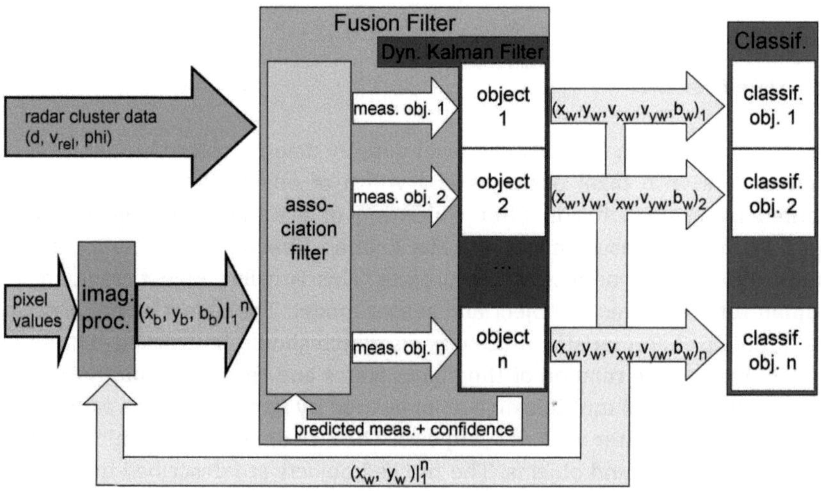

Fig. 1. Fusion concept

3.2 Classification

In our approach, the data sets which have to be classified and validated are image regions with tracked objects. The radar signature is not observed but could be included. To create the data base many traffic sequences with different objects under different environmental and exposure situations have to be evaluated. The interesting image regions are scaled to uniform size and stored in the data base. This is necessary to achieve a highly representative random sample. Based on these training data, two different classifiers, Multilayer Perceptron Networks (MLP) and Support Vector Machine (SVM) were trained and tested.

Multilayer perceptrons are a special kind of neural network with feed-forward topology. The neurons are grouped into several layers, the input layer, at least one hidden layer and the output layer. Each neuron in each layer has an adjustable weighted connection to each neuron in the next layer and so on. The image's pixels are assigned to the input generating an activation level for the neurons. The activation is propagated through the weighted connections and hidden layers up to the neurons in the output layer. The output neuron with the highest activation marks the "winning" class. To restrain false or insecure classifications, a threshold was determined by Receiver Operating Characteristic (ROC) evaluation. During training, weightings of the MLP are iteratively adjusted to minimize the networks mean squared error. Most of the training algorithms are based on the well known back propagation algorithm. When the training is finished the network must be able to classify untrained images correctly. In this state the network has a high generalization performance. [4] is a well-written introduction to the neural networks and the training algorithms, [6] provides the idea behind the perceptron model.

Support vector machines (SVM) are a new technique for data classification. They base on statistical learning theory. For classification with SVMs the image data is considered in a high dimensional feature space. Each image represents a data vector in this feature space. The support vector machine calculates a separating hyperplane between two different classes in the training set, for example the "car" class and the "garbage" class. The support vector algorithm calculates the hyperplane with the largest margin. The data vectors with the shortest distance to the separating hyperplane are called support vectors. To classify non-linear separable data we use an implicit mapping technique. General training of SVMs consists of the solution of a convex quadratic optimization problem. The resulting support vectors and coefficients like the Lagrange multiplier compose the description of the separating hyperplane. Fig. 2 shows exemplary the training vectors of two data classes with the resulting separating hyperplane in a not linear separable case. The classification of new, untrained images, results from calculating the side of the hyperplane on which the image data vector is located. If there's need to compare more than two classes, more than one SVM has to be utilised.

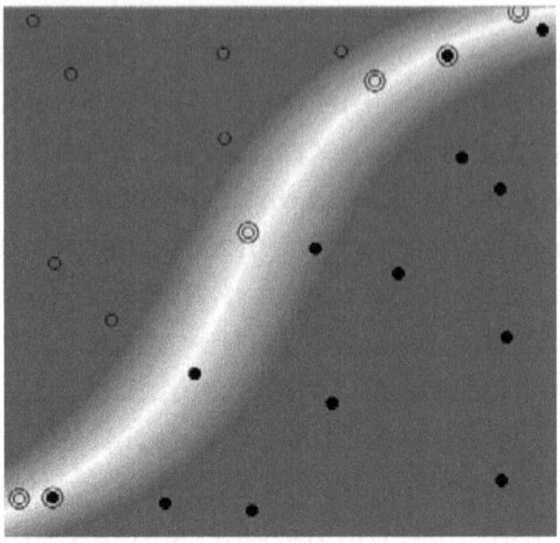

Fig. 2. Not linearly separable case: Circles are class 1, dots are class 2, marked circles are support vectors and the white curve is the separating hyperplane [7]

The fundamental problem of the Neural Network training methods is the missing global information concerning the error surface. Hence, it is possible that the minimized final training error is only a local minimum in the error surface. Thus, one can never be sure to have an optimal configuration of the weightings resulting in a real optimal classifier. So it is required to repeat the training procedure recurrently to obtain certainty. The advantage of Support Vector Machines compared to Neural Networks is, that they guarantee a global error minimum and with that an optimal classifier for the training data. This reduces training effort and improves generalization performance.

Fig. 3 shows the results of our fusion algorithm with a MLP classification for different relevant and irrelevant objects. Measured were a dummy with vehicle shape, radar measures of ghost targets and a special irrelevant object, a tin. The dummy is very similar to that of a real car camera and radar signature, so that this dummy should be detected. The figure shows, that the number of false detections can be reduced significantly between 5 m and 50 m distance to the objects, while most of the vehicles standing in the direction of traffic are matched to the vehicle class. Tests in real traffic conditions with different cars return similar results. The usage of support vector machines provides a similar performance.

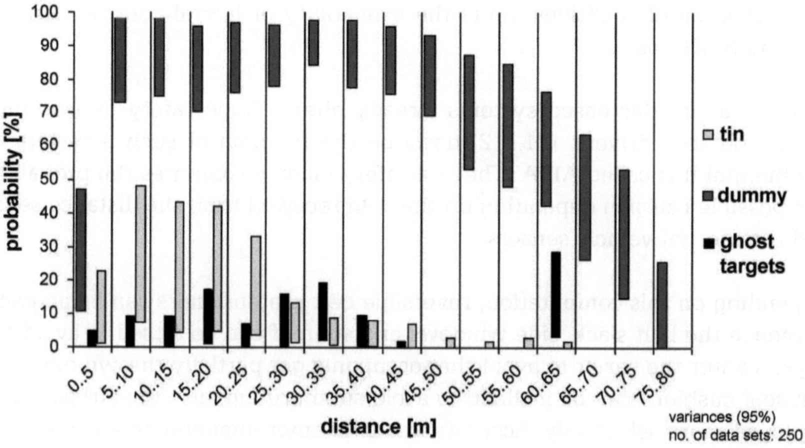

Fig. 3. Variance of probability statements for class "vehicle" in the cases of a dummy vehicle, ghost objects and a tin

To achieve best performance for both classification methods it is useful to pre-process the images to enhance the image quality. To improve the real time property it is useful to apply methods for decreasing the dimensionality of data like the Principal Component Analysis. [5] provides an introduction to this method, which can reduce dimensionality up to 90%.

4 Functional Improvements

Sensor fusion requires additional interfaces, resources and development effort, so it's significant if the additional comfort and safety justifies a fusion concept.

First of all a better object detection, in particular with regard to stationary objects, can improve existing comfort functions. Today's Adaptive Cruise Control (ACC) adjusts the speed of a vehicle depending on the distance and the relative speed to the next vehicle ahead based on a beam sensor only. [8] describes an ACC system that reduces false alarms and improves the lane assignment by the sensor fusion of radar and camera sensors. It is difficult to track a target vehicle at lower speeds and nearly impossible to accelerate the own vehicle after a standstill with sufficient reliability. The camera can create more confidence, to drive full speed range ACC e.g. in traffic jams. In addition, changes in situation can be detected faster and more reliable. Further, classical camera functions can be improved by high quality fusion based information.

The detection of roadsides limits the availability of lateral controls like lane keeping functions.

Also radar or lidar based systems already offer a huge safety improvement for unprepared drivers. [11, 12] describe the function of such a system. At Continental it is called APIA. There, a safety module computes the probability of a possible collision depending on the data received from the distance sensor and the inertial vehicle sensors.

Depending on this computation, reversible belt pretensioners can be activated to reduce the belt slack, side windows and sunroof can be closed to avoid that objects enter the car or to avoid that occupants get partially thrown out of the car, seat cushions can be inclined to avoid submarining, also the airbag can be deployed more selectively. Accurate and safe environmental data from sensor fusion reduces false alarms. Compared to simple beam sensor systems, more measures for pedestrian detection are obtained.

A further possible improvement is the control of active safety systems, especially the control of the brake system. Depending on beam sensor measurements and including the driver's behaviour regarding accelerator pedal travel, [11] describes the following active safety measures:

- Discrete warning, distance within the traffic situation.
- The brake system is preconditioned autonomously by prefilling. The measure shortens the response time and overcomes the system immanent free travel while the driver is still on the accelerator pedal. The driver is still able to overrule the system by actuating the accelerator pedal.
- When the driver releases the acceleration pedal, APIA actively applies the brakes up to a deceleration of 0.3 g to reduce the kinetic energy during the switching phase from accelerator to brake pedal.
- At this stage, the extended brake assist is activated. The deceleration, which would be needed in addition to the driver's braking for mitigating an impending crash, is added. As the brake system has already been prefilled in the former stages, maximum braking pressure can be applied rapidly, ensuring the shortest possible stopping distance.

Fig. 4 compares the pressure build-up of a normal unprepared driver without support, with brake assist function and with the support of a safety system based on a distance sensor. It describes the occurrence of a traffic hazard (danger) and the different stages of warning to the driver (Pre-Warning, Acute-Warning) dependent on time to collision (t_{ttc}). The driver's reaction (acc./ brake pedal contact) then initiates the deceleration support to the driver by the Electronic Stability Control (ESC) System. Activation thresholds of the regular

Brake Assist (BA) can be varied and the level of BA initiated deceleration (variable dep. on situation) is adjusted according to the traffic hazard.

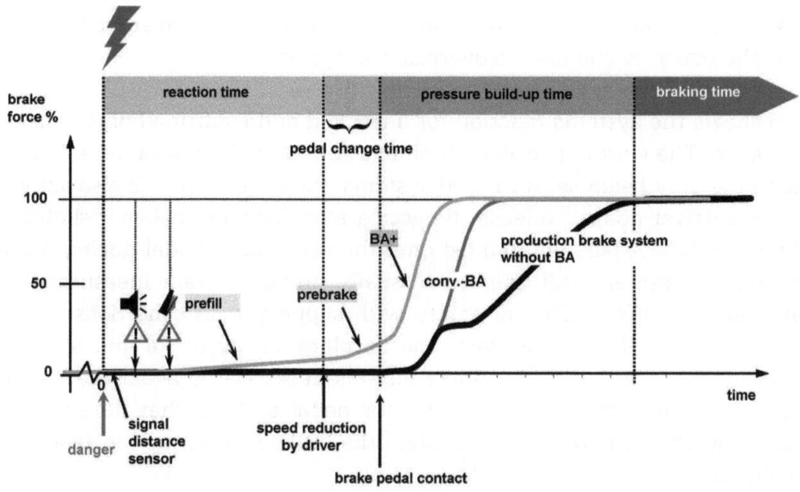

Fig. 4. Beam sensor based reduced stopping distance algorithm

False alarms might be generated by reflections on crash barriers if only a distance sensor is used. In addition it is difficult to deal with stationary objects. Only the movement of objects ensures the ability to distinguish them from the rest of the environment by tracking algorithms. To reduce the effect of unintended braking (e.g. rear impacts) because of ghost objects or irrelevant objects, the force of the brake interventions for moving objects has to be limited, too. Usually, the autonomously controlled deceleration doesn't exceed 0.5 g. To reduce the number of unintended brakings on ghost targets with a beam sensor based system, the driver behaviour has to be over weighted compared to the environmental information. The major need for plausibility with driver behaviour can lead to disadvantages compared to a fully automatic system. If the driver reacts late or wrong in these conventional systems, time for braking is lost.

Classification by camera reduces the detection probability of ghost targets to a minimum. Functions based on radar and camera should still consider the driver as far as reasonable, but the need to check the beam sensor measures for plausibility with the driver behaviour, to eliminate reactions on ghost targets, is reduced. For the system described in the following this enables a higher availability of the function, stronger deceleration and also a more sophisticated intervention strategy. The system contains the following improvements:

- inclusion of classified stationary objects in the control strategy
- higher deceleration without brake pedal contact

- disregard, if the driver's foot slightly falls onto the accelerator pedal
- higher maximal deceleration when releasing the acc. pedal without brake pedal contact
- prebrake and torque reduction independent of the driver's behavior
- the driver is still able to overrule the system

Fig. 5 shows the systems reaction for a tracked and confirmed object on collision course. The warning strategy is similar to that without a camera. In difference to the only beam sensor based systems, the deceleration is also activated when the driver doesn't release the accelerator pedal or pushes it slightly. So if the driver's foot falls unintended onto the accelerator pedal during a brake situation, the system still operates. Usually proper friction measurement is unavailable. To offer sufficient safety with slippery road conditions, without this knowledge, the system starts the deceleration by brake and a reduced engine torque, if at least 40% brake force is necessary to avoid the collision. Only if the driver pushes the accelerator pedal so hard that he apparently wants to overrule the brake intervention the brake intervention is deactivated (see Fig. 6).

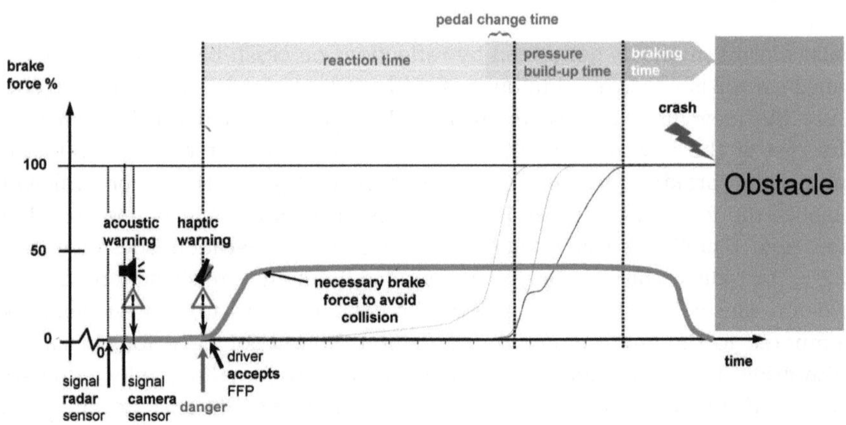

Fig. 5. Camera validated emergency braking, driver accepts force-feed-back-pedal (FFP)-warning

The brake force is deactivated, until the driver recognizes the danger and releases the accelerator pedal. In this situation a higher deceleration is necessary to avoid a crash. If an obstacle occurs late or the detection range of the sensor isn't sufficient to stop the vehicle with this medium deceleration, the system increases the deceleration independent of a brake pedal contact up to

0.7 g. In case of a brake pedal contact the brake force is increased up to 100% if the road conditions permit.

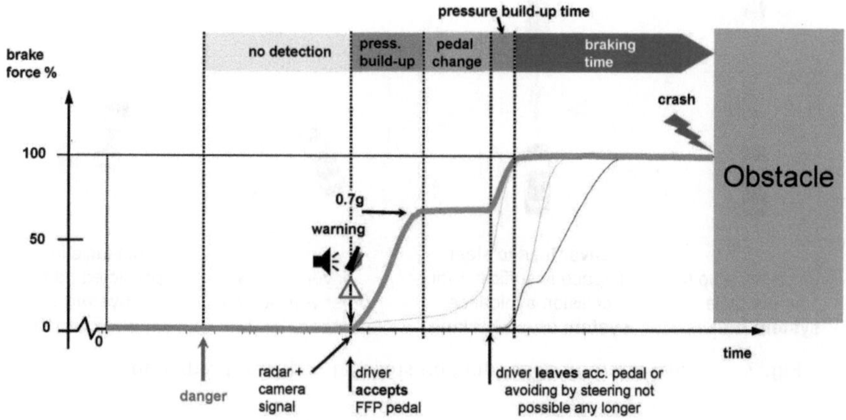

Fig. 6. Camera validated emergency braking, driver accepts FFP-warning very late and later leaves acceleration pedal

In some emergency situations, particularly at higher speed, driving around an obstacle is the better solution to avoid a collision. There are different possibilities to realize this in the intervention strategy. [13] describes a fully autonomous steering intervention for crash avoidance. Even if this autonomous decision is far away from serial production, as a first step it is necessary to include the driver's steering reaction in the brake strategy. It would be possible to keep the brake force as long as the vehicle is on collision course and to rely on the remaining steering abilities in case of ABS intervention. One problem is that the vehicle looses steering agility with ABS intervention; the other problem is the reduced driver's feedback. If the brake force is reduced to a minimum or is completely switched off when the steering angle promises crash avoidance, the driver receives feedback that his strategy is correct. The system releases the brake force as soon as the current steering angle can avoid a crash as described in Fig. 7. If the remaining speed and distance to the obstacle becomes critical again, the system repeats the build up of the brake pressure.

Fig. 7. Brake strategy depending on steering and path prediction

The behaviour on classified objects is described, but how to deal with unconfirmed objects? There are possibilities to evaluate unconfirmed objects as ghost objects as described in [2], but independently a radar-only based algorithm with driver confirmation should be computed as fall back solution and redundancy. So failure of camera or video processing can never reduce the performance compared to a beam sensor only based system.

5 Results

To test the camera and radar fused system the vehicle was driven onto a stationary obstacle with several velocities between 20 and 110 km/h without any driver reaction and sufficient road friction. Fig. 8 plots the remaining speeds in front of the object depending at different approaching velocities. During the test drive it was impossible to measure the speed at the collision itself. Therefore the measures of the remaining speed 2 m in front of the obstacle are considered. With a maximum deceleration of 7 m/s^2 it is still possible to reach a standstill. All crashes up to 90 km/h were avoided, if the obstacle was detected. The vehicles detection rate at 90 km/h is about 90%. Below this speed down to 40 km/h the detection rate rises up to 98%. If this detection performance could be accomplished in real traffic conditions, rear end collisions with relative speeds up to 90 km/h could be avoided for the most part. If the obstacle is moving, the maximum speed of the host vehicle increases depending on the obstacle's velocity. But this is not considered here.

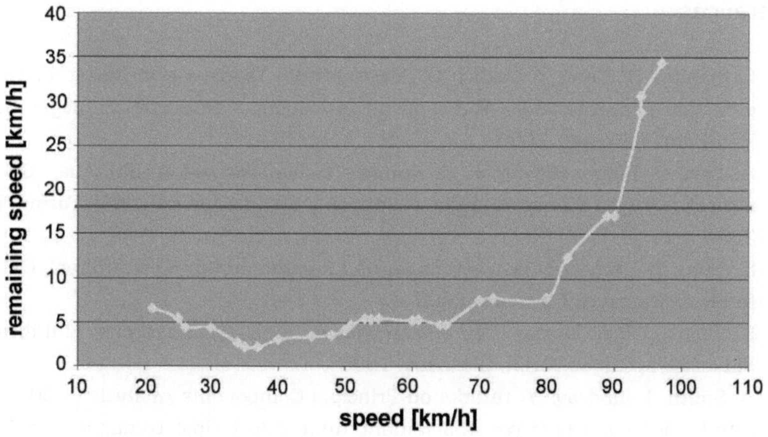

Fig. 8. remaining speed 2 m in front of obstacle depending on approaching velocity

In the case of higher speeds the system could not avoid the crash but is still able to reduce the severity of the impact. In addition the time between the beginning of the brake intervention and the crash is longer at high relative speeds, so that the driver has plenty of time to recognize the situation and to avoid the possible collision by a steering intervention. This case was tested with speeds up to 130 km/h and it was always possible to pass the obstacle after a reaction time of 1 second.

6 Summary

Even with high performance radar systems, a fusion and classification by camera improves the reliability and accuracy of the object detection, particularly for stationary objects like a traffic jam end. Multilayer perceptrons and support vector machines offer a good possibility to classify objects. In comparison with multilayer perceptrons support vector machines guarantee limited training effort.

With the additional reliability for classified objects an improvement of active brake functions is possible. The presented system brakes automatically on moving and stationary objects. The reduced need to check the beam sensor information for plausibility with driver behaviour can generate more safety.

References

[1] L. Bombini, P. Cerri, P. Medici, G. Alessandretti, "Radar-vision fusion for vehicle detection", International Workshop on Intelligent Transportation, pages 65-70, Hamburg, Germany, 2006

[2] A. Sole, O. Mano, G.P Stein, H. Kumon, Y. Tamatsu and A. Shashua, "Solid or not solid: Vision for radar target validation", Symposium (IV2004), Parma, Italy, 2004

[3] S. Heinrich, "Sensor-Fusion von Radar und Kamera mittels Kalmanfilter", CTI-conference, Stuttgart, Germany, 2005

[4] R. Rojas, "Theorie der Neuronalen Netze: Eine systematische Einführung", Publisher Springer Berlin, Germany, 1996

[5] A. Smith, I. Lindsay, "A tutorial on Principal Components Analysis", 2002, URL: http://csnet.otago.ac.nz/cosc453/student_tutorials/principal_components.pdf

[6] F. Rosenblatt, "The perceptron: a probabilistic model for information storage and organization in the brain", Psychological Review, Vol. 65, 1958

[7] C. J. C. Burges, "A Tutorial on Support Vector Machines for Pattern Recognition",Bell Laboratories, Lucent Technologies, 1999

[8] U. Hofmann, A. Rieder, E.D. Dickmanns, "Radar and Vision Data Fusion for Hybrid Adaptive Cruise Control on Highways", Machine Vision and Applications, Publisher Springer Berlin / Heidelberg, Germany, 2003

[9] U. Hofmann, A. Rieder and E. D. Dickmanns, "EMS-Vision: Application to Hybrid Adaptive Cruise Control", Proceedings of the IEEE Intelligent Vehicles Symposium, Dearborn, MI, USA, 2000

[10] A. Gern, U. Franke, P. Levi, " Advanced lane recognition fusing vision and radar", Proc. IEEE Intelligent Vehicle Symposium 2000, page 45-51, Detroit, USA, 2000

[11] R. Adomat, G. Geduld, M. Schamberger, Jürgen Diebold, M. Klug, "Intelligent Braking: The Seeing Car Improves Safety on the Road", Advanced Microsystems for Automotive Applications, Publisher Springer Berlin / Heidelberg, Germany, 2005

[12] J. Diebold, "Active Safety Systems – The Home for Global Chassis Control", Convergence International Congress and Exposition on Transportation Electronics, Detroit, MI, USA, 2006

[13] E. Bender, M. Darms, M. Schorn, U. Staehlin, R. Isermann, " Das Antikollisionssystem PRORETA – auf dem Weg zum unfallvermeidenden Fahrzeug", Automobiltechnische Zeitung ATZ 02, Vieweg Verlag, Wiesbaden, Germany, 2007

Dr.-Ing. Stefan Lüke

Continental Automotive Systems

Guerickestraße 7

60488 Frankfurt am Main

stefan.lueke@contiautomotive.com

Dipl.-Ing. Matthias Komar
Continental Automotive Systems
Kemptener Strasse 99
88131 Lindau
matthias.komar@contiautomotive.com

Dipl.-Ing. Matthias Strauss
Continental Automotive Systems
Guerickestraße 7
60488 Frankfurt am Main
matthias.strauss@contiautomotive.com

Keywords: ADAS, driver assistance systems, sensor fusion, kalman filter, classification, multilayer perceptron network, support vector machine, emergency braking, force feedback pedal

Networking Sensors and Actuators for a New Active Headrest

J. Murgoitio, M. Ferros, A. Goti, M. Larburu, T. Rodríguez, Robotiker-Tecnalia

Abstract

Tecnalia Automoción is working on new concepts for car cockpit components based on "Ambient Intelligence" (AmI) principles. Related to this concept, Tecnalia Automoción has developed a prototype of a sensorized active headrest to be placed maintaining desired horizontal and vertical safety distances to head and without pyrotechnical actuators which can cause injuries to the user. With the implementation of AmI based solutions like described above, the number of car sensors and actuators will grow up drastically, and more efficient control architectures will be needed. For this, Tecnalia Automoción is designing a networked solution with smart sensors and actuators integrating the IEEE 1451 standard group, initially to be applied to active headrest prototype. IEEE 1451 is a group of seven standards, some of them in revision phase, about smart transducers interface for sensors and actuators that they will let us to have more features in sensor-actuator side.

1 Introduction

Nowadays, a medium car has between 50 and 100 sensors and this number is growing up. It means that solutions about networking sensors and more intelligent sensors (more features in sensor side) are being considered as strategic ways in the future, not only for sensors but for actuators too as elements to execute actions within a distributed control system. Related to previously mentioned, Robotiker is researching about intelligent sensor networks and his availability and application in car systems and devices. All mentioned sensors and actuators would be integrated within a networked system taking into account the IEEE 1451 standard for a "plug&play" system.

2 AmI Solution: Active Headrest

The new active headrest developed by Robotiker Tecnalia detects the position of the user's head using the measured information by contact-less devices:

two infrared sensors placed in the headrest (ultrasound sensors are also being evaluated as an alternative for occupant´s head detection). One of them is a presence detecting sensor and the other one is an analogical sensor to measure the distance to an object. With them, we can find where the edge of the head is and we can know the distance from the headrest to the head. The system adjusts the headrest to the optimal safety position using two independent motion controls for the horizontal and vertical directions. A clarifying scheme is shown in Fig. 1:

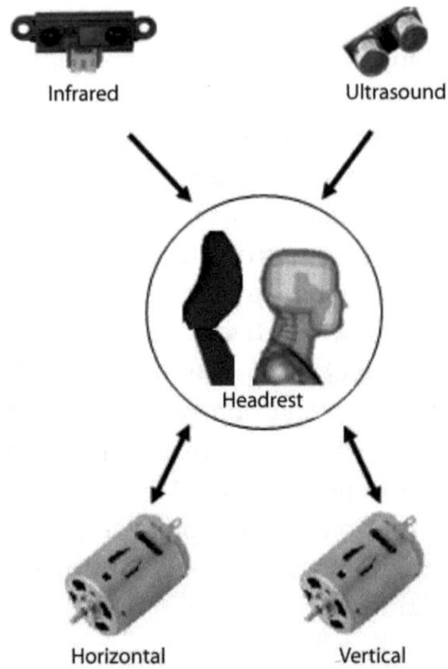

Fig. 1. Active headrest

This system controls two non-interpolated axes within a headrest system to maintain safety distances between the headrest and the car occupant´s head: the horizontal distance from headrest to head and the vertical distance from the top of the head to top of headrest.

Nowadays, a group of two infrared sensors in the headrest device to collect information about the car-occupant position, and two more different sensors to know the current head-rest position are working all together in the system. In addition, two actuators are commanded to carry out the application goal of maintaining these two safety distances. This control is done in a continuous way at regular intervals of a programmable time.

The described application will take care over the recommended horizontal and vertical safety distances, according to recognized prevention guides and standards.

3 IEEE1451: Networking Sensors and Actuators

IEEE 1451 is a family of smart transducer interface standard to make easier for transducer manufacturers to develop smart devices and to interface those devices to networks, systems, and instruments by incorporating existing sensor and networking technologies. The goal of 1451 is to allow the access of transducer data through a common set of interfaces whether the transducers are connected to systems or networks via a wired or wireless means.

IEEE 1451 is a group of seven standards, some of them in revision phase:

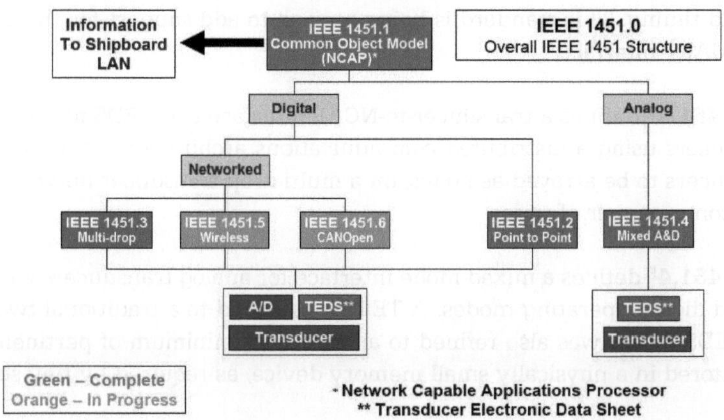

Fig. 2. Family of IEEE 1451 (NAVSEA)

The **family of IEEE 1451** standards are sponsored by the IEEE Instrumentation and Measurement Society's Sensor Technology Technical Committee and describes a set of open, common, network-independent communication interfaces for connecting transducers (sensors or actuators) to microprocessors, instrumentation systems, and control/field networks. (^{D} - it is being developed; ^{P} - it is in publication and can be acquired at the IEEE.org web site)

IEEE P1451.0^{D} provides a uniform set of commands, common operations, and TEDS for the family of IEEE 1451 smart transducer standards. This command set lets to access any sensors or actuators in the 1451-based networks. This standard will be used to assure uniformity within the family of IEEE 1451.x interface standards.

IEEE 1451.1[P] defines a common object model describing the behaviour of smart transducers. It defines the communication models used for the standard, which included the client-server and publish-subscribe models. Application software running in the NCAP based on IEEE 1451 communicated with transducers through any IEEE 1451.X physical layer standards as required for a particular application.

IEEE 1451.2[P] defines a transducers-to-NCAP interface and TEDS for a point-to-point configuration. Transducers are part of a Transducer Interface Module (TIM). The original standard describes a communication layer based on enhance SPI (Serial Peripheral Interface) with additional HW lines for flow control and timing. This standard is being revised to add support for the popular serial UART interface.

IEEE 1451.3[P] defines a transducer-to-NCAP interface and TEDS for multi-drop transducers using a distributed communications architecture. It allows many transducers to be arrayed as nodes, on a multi-drop transducer network, sharing a common pair of wires.

IEEE 1451.4[P] defines a mixed-mode interface for analog transducers with analog and digital operating modes. A TEDS was added to a traditional two-wire. The TEDS model was also refined to allow a bare minimum of pertinent data to be stored in a physically small memory device, as required by tiny sensors.

IEEE P1451.5[D] defines a transducer-to-NCAP interface and TEDS for wireless transducers. Wireless communication protocol standards such as 802.11 (WiFi), 802.15.1 (Bluetooth), 802.15.4 (ZigBee) are being considered as some of the physical interfaces for IEEE P1451.5.

IEEE P1451.6[D] defines a transducer-to-NCAP interface and TEDS using the high-speed CANopen network interface. It defines a mapping of the 1451 TEDS to the CANopen dictionary entries as well as communication messages, process data, configuration parameter, and diagnosis information.

Definition of Transducer Electronic Data Sheets (TEDS) is the key feature of this family of standards and it would be a memory device attached to the transducer having information like transducer identification, calibration, correction data, measurement range, manufacture-related information, and so on.

4 IEEE 1451 for the Active Headrest

Sensors and actuators in our active headrest system will be integrated within a networked system taking into account the mentioned IEEE 1451 standard for a "plug & play" system as it is shown in the following figure:

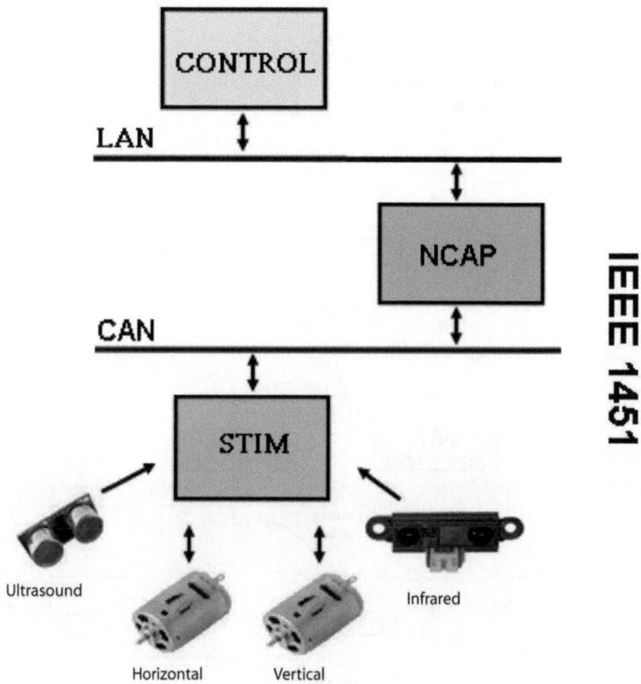

Fig. 3. Sensor / actuator network: general overview

According to the IEEE 1451 standard the active headrest will be integrated in a system with one NCAP (Network Capable Application Processor) and one STIM (Smart Transducer Interface Module). As it's shown in the figure, the STIM will collect information from four different **sensors** (ultrasound, infrared, horizontal and vertical). In the same way, the STIM will be connected to the two **actuators** of the head-rest system (horizontal and vertical). So, all sensors' data will be accessible from NCAP to the selected LAN (Local Area Network) where "control" will be able to use all this information, and according the control system actuate over the two actuators (horizontal and vertical) through NCAP and STIM components.

4.1 CONTROL

An overview of the control carried out in our system is shown in the following figure:

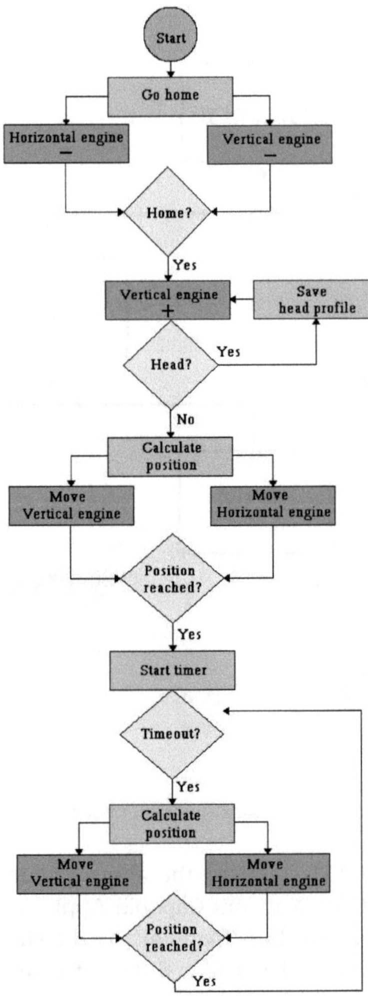

Fig. 4. Control

Nowadays, this "control system" starts when the "key" of the car is switched on and the "active headrest" system is selected. Then, "horizontal" and "vertical" engines move until the "home" position is reached, it is the lowest position for vertical axe and far away from the head for horizontal axe. At this point, only the vertical engine starts moving from lowest position (home) to highest position while "head" is detected by an infrared sensor. During this movement a profile of the head is generated taking into account a different sensor sending information about the distance from the headrest to the car-driver head. When this head-profile is obtained the control system calculates the proper position for the headrest taking into account some programmable parameters like "horizontal" and "vertical" distances to maintain from the driver head to the active headrest, and the two engines are independently moved to reach the calculated security position. After that, a programmable timer is activated to recalculate the security position for the headrest every "timeout" to move "horizontal" and "vertical" engines an keep the proper security position.

4.2 NCAP

As it is shown in the following figure, NCAP will interchange data between a CAN network and a particular LAN:

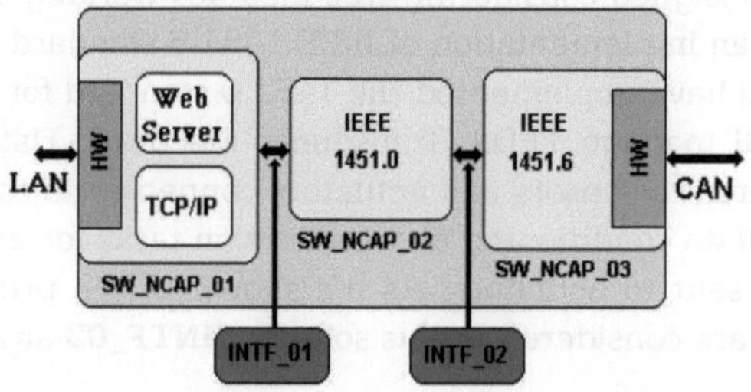

Fig. 5. NCAP (Network Capable Applications Processor)

Tree general modules will work together to carry out this task. **SW_NCAP_01** has been designed to connect all sensors and actuators with "Control" by means of a particular LAN. Nowadays, TCP/IP protocol and a web-server mechanism is being considered as a demonstrator. **SW_NCAP_02** will implement the 1451.0 standard to provides a set of common commands and operations, to access any sensors or actuators in the 1451-based networks assuring uniformity within the family of IEEE 1451.x. Finally, **SW_NCAP_03** will develop the 1451.6 standard to communicate with a CAN network where several STIM modules can be connected. Two interfaces are considered in this solution as it can be seen in the previous figure **(INTF_01** and **INTF_02**).

4.3 STIM

A general overview of the STIM is shown in the following figure:

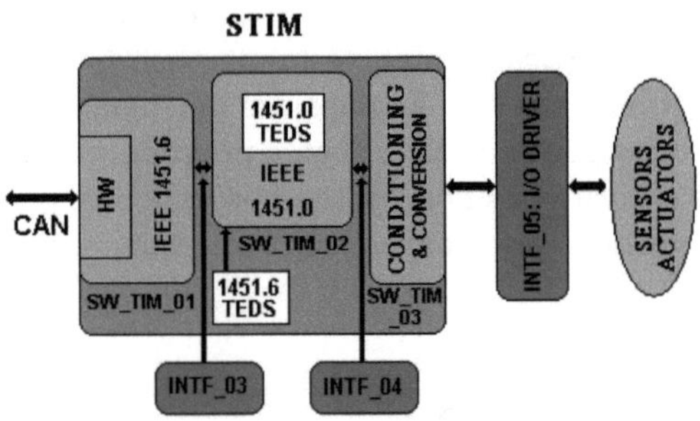

Fig. 6. STIM (Smart Transducer Interface Module)

STIM has been designed considering tree modules working together too. **SW_TIM_01** will be an implementation of IEEE 1451.6 standard but in STIM side. **SW_TIM_02** will have implemented the 1451.0 standard for the sensor/actuator side and will manage TEDS (Transducer Electronic Data Sheet) with all information related to sensors and actuators connected to this STIM. Finally, **SW_TIM_03** will do conditioning and conversion tasks for data received from sensors or data sent to actuators. As it's shown in the previous figure, two more interfaces are considered in this solution (**INTF_03** and **INTF_04**).

5 Conclusions

Related to "AmI" concept, networking sensors and more intelligence in sensor side, are being considered as strategic ways in the future. In this way, several applications within confort and security systems for transport are being developed in Robotiker-Tecnalia foundation taking into account new standards.

IEEE 1451 family standard:

- Define network-neutral and vendor-independent transducer interfaces, and standardized transducer electronic data sheets (TEDS) that contain manufacture-related sensor data.
- Support a general model for transducer data, control, timing, configuration and calibration.
- Eliminate manual entering of data and system configuration steps, ultimately achieving "Plug&Play".
- Allow transducers (sensors and actuators) to be installed, upgraded, replaced or moved with minimum effort.
- Enable users to access wired or wireless sensor data and information seamlessly from a host system or network anywhere.

Benefits using this type of solution would be:

- For end users will reduce setup, maintenance, and overall life cycle costs with easy-to-use, plug-and-play features. IEEE 1451 will assist users in identifying, calibrating, analyzing, locating, and monitoring IEEE 1451-compliant transducers.
- For manufacturers of sensors and actuators will benefit from the use of a single standard for "network neutral" hardware and software interfaces. In addition, IEEE 1451 calibration mechanisms will help manufacturers develop better components.
- For transducer system integrators will reduce expenditures on transducer-to-network interface development efforts. System integrators will also be able to leverage existing resources by modifying existing IEEE 1451 reference implementations.

References

[1] Gorman, Bryan. Towards a Standards-Based Framework for Interoperable CBRN Sensor Networks . [Conference]. Hampton-Virginia: SensorsGov Expo & Conference, USA, December 2005.

[2] Wiczer, James; Lee, Kang. A Unifying Standard for Interfacing Tansducers to Networks – IEEE-1451.0 . [Workshop]. Lviv: IEEE International Workshop on Intelligent Data Acquisition and Advanced Computing Systems: Technology & Applications, Ukraine, 8-10 September.

[3] Lee, Kang. Synopsis of IEEE 1451 . [Conference]. Chicago: Sensors Conference/Expo 2005, USA, June 2005.

[4] RCAR: Research Council for Automobile Repairs. A Procedure for Evaluating Motor Vehicle Head Restraints (Issue 2). February 2001. Web: http://www.rcar.org/papers/rcar.pdf.

[5] Michael, Kleinberger; Emily, Sun; James, Saunders; Zaifei, Zhou. Effects of head restraint position on neck injury in rear impact. Traffic Safety and Auto Engineering Stream of the Whiplash-Associated Disorders World Congress, Vancouver, Canada, 7-l 1 February 1999.
[6] Patent: Apparatus for determining the localization of a head of an occupant in the presence of objects that obscure the head. US 6088640.
[7] Patent: Automatic headrest adjustment using sensors. GB 2383530.
[8] Patent: Headrest for an automobile seat. EP 1270316.
[9] Patent: Method and device for headrest control. JP 3221005.

J. Murgoitio, M. Ferros, A. Goti, M. Larburu, T. Rodríguez
Parque Tecnológico, Edificio 202
E-48170 ZAMUDIO – Bizkaia
Spain
murgoiti@robotiker.es

Keywords: Ambient intelligence, sensor, actuator, sensor network, IEEE1451, active headrest.

Classification of Road Conditions – to Improve Safety

J. Casselgren, M. Sjödahl, University of Technology Luleå
S. Woxneryd, M. Sanfridsson, Volvo Technology Corporation

Abstract

Measuring the road condition in front of a vehicle could prevent accidents and make technologies like electronic stability control (ESP) more efficient. By making three investigations of the classifications of the four road conditions dry asphalt, asphalt covered with water, ice and snow the possibility of a preview sensor is exploited. By measuring the reflectance from the different surfaces with a halogen light and an actual sensor (Road eye) in a laboratory surroundings the advantage and disadvantage are revealed. The sensor is also mounted in a Volvo truck for real-life condition measurements.

1 Introduction

Slippery roads show high statistics of accidents [1]. If the low friction number could be measured by a preview sensor this could work both as a warning for the driver and incorporated in the technical systems of the car to make them work more efficient. The distinction in reflectance of different road conditions as asphalt and asphalt covered with water, ice and snow can be exploited in a sensor for classification of the road surface ahead of a vehicle.

The purpose of this investigation is to describe the reflectance characteristics of the four road surfaces for the wavelength around 1300 nm and 1550 nm. Both with unpolarized light with several different angles and with polarized light with the angle and distance specified of the constructor of the Road eye (Swedish patent nr 9904665-8) sensor. The two measurements are compared to guarantee the robustness of the method used in the sensor. The sensor was also tested in real-life condition on test tracks in Arjeplog, in the north of Sweden. The experimental setup for the experiments is explained in section 2. In section 3 the result is presented and discussed. The paper is ended with conclusions.

2 Experimental Setup

The three experimental setups that were used allowed us to investigate the reflectance effects of four road surfaces, dry asphalt and asphalt covered with water, ice and snow. Two experiments were carried out in a laboratory environment in order to control the depths and purity of the layers covering the asphalt. The first experiment the reflectance of the surfaces, illuminated with a halogen light (unpolarized light), was measured with a spectrometer this is explained in more detail in section 2.1. In the second experiment the Road eye Sensor that consist of laserdiodes for illumination (polarized light) and photodiodes for detection was applied on the same surfaces as in the first experiment. The approach of the experiment is explained in section 2.2.1. Section 2.2.2 describes the third setup were the sensor was mounted on a Volvo truck to investigate the sensor in working order.

Fig. 1. a) The half sphere with the light source positioned at 90°; b) the light source with the lens mounted at the end; c) the piece of ABT 11 asphalt; d) the spectrometer and the modifier numbered with (1)

1.1 Laboratory Measurements Halogen Illumination

Fig. 1 shows the first experimental setup: Fig. 1a consisting of a halogen light for illumination, mounted together with a lens with +150 mm focal length,

Fig. 1b giving a 50 mm light spot on the asphalt, Fig. 1c positioned in the centre of the (poly methyl methacrylate) half sphere. The asphalt was a dense bitumen tar pulp (ABT 11) mixed with crushed stone components with the largest size of 11 mm. The crushed stone changes depending on local assets of stone material which also can alter the spectrum of the asphalt. ABT 11 was chosen since it is common asphalt on Swedish roads representing about 80% of all hot manufactured asphalt coatings in the country.

The half sphere was drilled with holes with an angular resolution of 10° from 0° to 70° vertically, where 0° is perpendicular to the asphalt. Horizontally it was drilled in entire circles around the sphere also with an angular resolution of 10° pointing at a spot at the centre of the half sphere. The holes are 3 mm in diameter to allow insertion of the modified spectrometer (Fig. 1d) so it is perpendicular to the sphere's surface and measures on the same point at the centre of the sphere. The spectrometer is a NIR 1.7–Spectrometer, from Boehiringer Ingelheim microParts GmbH, which measures in the range of 1100-1700 nm with a numerical aperture of 0.22. The spectrometer is modified with an extension ((1) in Fig. 1d) with a numerical aperture of 1.5 mm and a length of 18 mm to have a field of view of 20 mm on the asphalt. Note that the measured spot was smaller than the light spot to ensure that the spectrometer measurements were restricted to within the illumination spot.

The spectrometer was set to measure the percentage of the reference light that is reflected from the surface. The reference light was measured strait on the illumination at a distance twice the radius. To get a good assessment of the layers reflection it was important that the experiments were carried out in the same way for all measurement cases. For that reason a lift was placed under the asphalt to enable the top surface always to be at radius distant from the sphere. Four different conditions were investigated; dry asphalt and asphalt covered with water, ice and snow. Altering the water depth was achieved by filling or emptying the container holding the asphalt sample. The level of water was measured at the same point on the asphalt which was the zero level reference point for the rough surface of the asphalt. The water and snow depth was measured with a slide-calliper. The same point was also used for the ice as zero level reference point, the ice thickness was measured with a micrometer. The snow grains were filtered through two filters to achieve a grain size between 1 and 2 mm, most of the grains were of these sizes. A sieve was used to apply the snow on the asphalt to get an even distribution. In this case an assessment was made of the snow depth.

To preserve the ice and snow, the measurements were performed in a climate room with temperature -10°C. The water and ice was measured for five different layers, moist or ice crystals and a thickness of 1-4 mm, the snow was only measured with thicknesses of 1 and 3 mm because of problems in obtaining a flat surface. The 3 mm snow surface was then pressed with an ice hockey puck and a weight of 3.4 kg. In this investigation only the back scattered light was analyzed for all angles and direction of illumination between 0°-70° vertically. In the case where the light position and measuring point were the same the measurement was done with a small change of the horizontal angle of 2-3°, due to the width of the light source. The results were analyzed in Matlab.

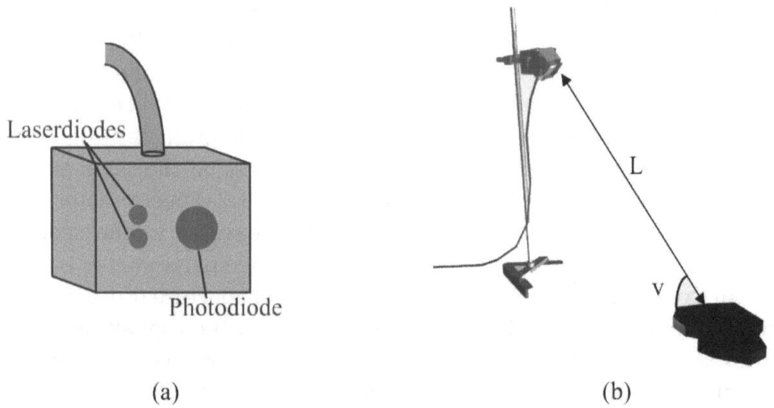

Fig. 2. a) Schematic drawing of the Road eye sensor; b) setup for laboratory measurements with the sensor

2.2 Road Eye Sensor

The Road eye (Swedish Patent) sensor consists of two laserdiodes for illumination and a photo diode for detection mounted as shown in the schematic drawing in Fig. 2a. The two lasers works at 1300 nm and 1550 nm. The wavelengths agree with the optimal wavelength for classification as shown in [2]. To focus the reflections on the photo diode a lens is placed in front. The sensor operates on 12 V and has a four value output. The two first values are the reflected optical power of the lasers respectively. The third value is the ratio between the two wavelengths, by coincidence this value can be compared to a friction coefficient. Fourth value is the square root mean of the reflectance of the two wavelengths. The sampling rate of the sensor is 30 measurements per second.

2.2.1 Laboratory Measurements

The laboratory setup of the sensor is seen in Fig. 2b where the distance L varied from 650 mm to 800 mm and with an angle (v) of 45°. The asphalt used in the first experiment is also used in the second experiment. The preparation of the layers was carried out in the same way for both experiments but the depths varied. The depth of water was moist, 1 mm 2 mm and 3 mm and the ice depths was 0.6 mm, 1.5 mm and 3 mm. The snow depths were 1 mm, 3 mm and 3 mm compressed in the same way explained in Section 2.1. The sensor was controlled and supervised with Labview and the measurements consisted of 256 to 1000 values for both wavelengths.

The advantage with the Road eye sensor was that measurements could be executed for surfaces with a fast change like ice and water. Therefore the 3 mm ice was also investigated covered with 1 and 3 mm water. Also two frost depths were applied on the asphalt and investigated but the depths could not be measured accurately it was estimated to 1 mm and 3 mm.

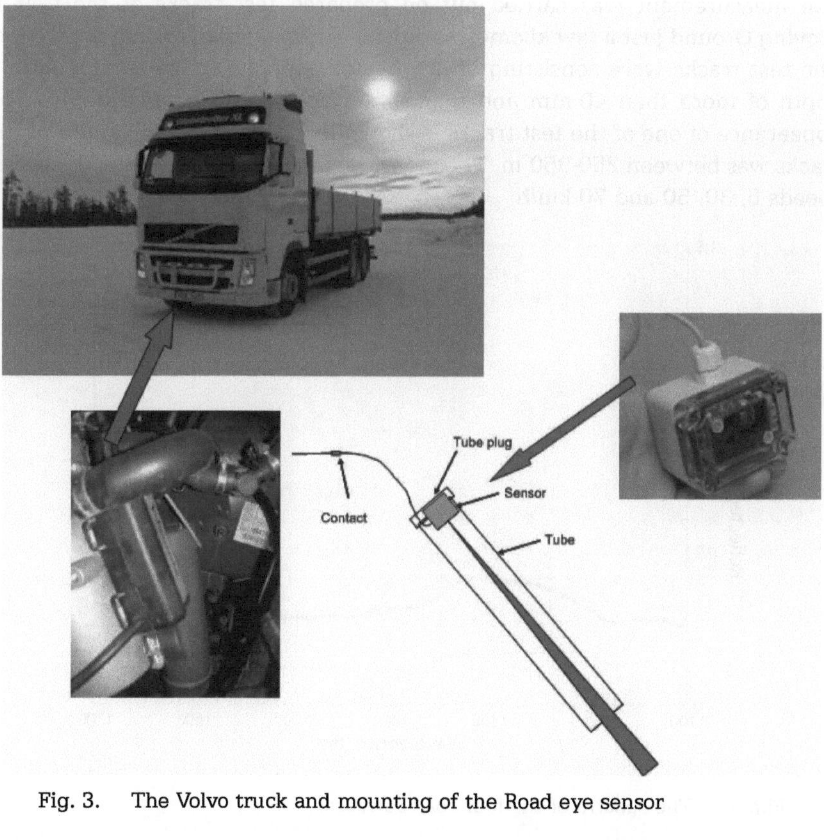

Fig. 3. The Volvo truck and mounting of the Road eye sensor

2.2.2 Real-Life Condition Measurements

The Road eye sensor was mounted in a Volvo truck. Considering the design of the Road eye sensor, it was essential to keep it from dirt during runtime; the sensor was therefore placed in a plugged tube. The construction was mounted in the wheel house at the height of 640 mm from the asphalt with a ~25° angle. It was mounted so that the lasers were pointing just under the bumper as in Fig. 3. The sensor was connected to a computer where the measurements were saved.

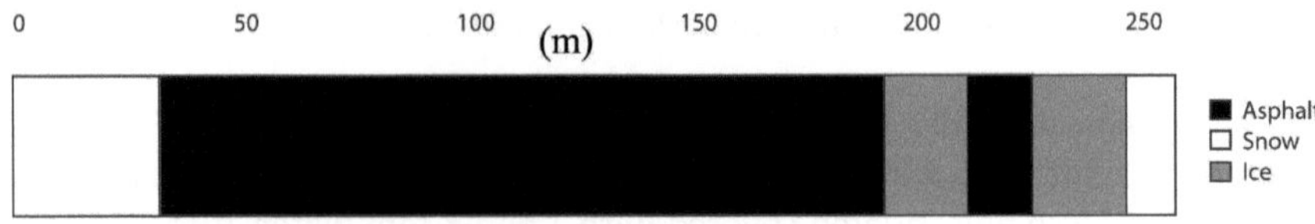

Fig. 4. Example of test track layout in Arjeplog

The measurement was carried out on prepared test tracks at the Colmis Proving Ground just a few kilometres outside Arjeplog in the north of Sweden. The test tracks were consisting of dry heated asphalt, an ice surface with a depth of more than 40 mm and a snow surface with the same depth. The appearance of one of the test tracks is shown in Fig. 4. The total length of the tracks was between 250-350 m. The measurements was executed with several speeds 5, 30, 50 and 70 km/h.

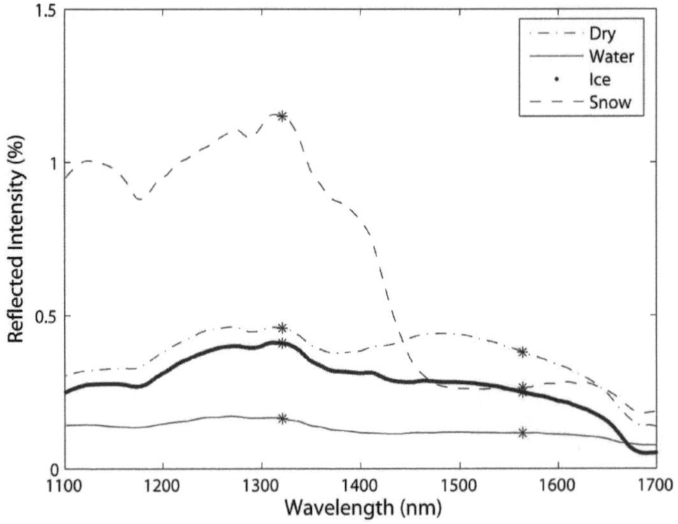

Fig. 5. The spectra of the four road surfaces

3 Results and Discussion

The halogen light is continuous over the spectrum measured by the spectrometer. By averaging all the measured spectra for a surface, the computation results in an estimate of the reflection for that surface. Shown in Fig. 5 are the spectra for the dry, wet, icy and snowy asphalt. The conclusion of Fig. 5 is that the four surfaces reflect light differently which makes the classification possible. The wavelengths used in the Road eye are marked with *. The wavelength 1300 nm is the reference wavelength and 1500 nm is the absorbing wavelength. The data collected for the two wavelengths are absolute values that could change depending on the asphalt but the spectrum are unchanged for all the surfaces. [2]

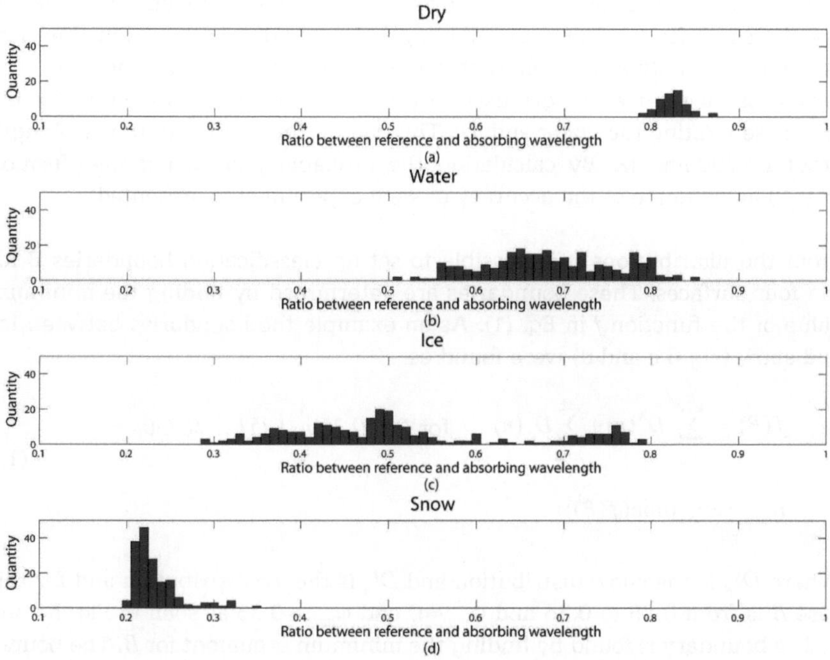

Fig. 6. The ratios distributions of the laboratory halogen light measurements

This variation in level for the measurements is avoided by computing the ratio for the two wavelengths for classification. Note that this also results in a sensor robust for intensity changes, which can be caused by a number of factors (changes of distance to the measuring point, a bump, or changes in the reflection as for new and old asphalt or dirty sensors). The assumption done for this solution is that the intensity changes the same over all wavelengths.

The ratio of the two wavelengths is thus insensitive to intensity changes. However, when the road surface changes the ratio will indeed be altered. For the sensor the ratio is calculated directly in the sensor and for the laboratory measurements the analyses were done in Matlab for the same wavelengths used in the Road eye sensor.

3.2 Laboratory Measurements Halogen Illumination

In this series of measurements the illumination was a halogen light and the detector the spectrometer. The ratio of the data are summarized in Fig. 6 a-d. The four different surface distributions are represented of for the dry 56 measurements, water and ice 280 measurements and snow 168 measurements. The distributions show that the surface can be classified but some values are the same for two surfaces. The dry and snow distributions are the most defined, with only a few values shared with water and ice respectively. The weakness for the wavelength used in this investigation is, as seen in Fig. 6b) and c), separating the water and ice. This can be done with a third wavelength around 1700 nm. [2] By calculating the probability of wrong classification (P_{wrong}) an estimate of the accuracy of each experiment is presented.

From the distributions it is possible to set up classification boundaries B for the four surfaces. These boundaries are determined by finding the minimum value of the function f in Eq. (1). As an example the boundaries between ice and snow (Fig 6 c and d) were found as:

$$f(B) = \sum_{n=\ddot{o}_{\min}}^{B} D_A^S(n) + \sum_{n=B}^{\ddot{o}_{\min}} D_A^I(n) \qquad \text{for } B = 0.250, 0.251 \ldots 0.350 \tag{1}$$

$$B_{\min} = \arg_B(\min(f(B)))$$

Where $D^S{}_A$ is the snow distribution and $D^I{}_A$ is the ice distribution and for this case B is from 0.25 to 0.35 and $\varphi_{min}=0$ and $\varphi_{max}=0.35$ as seen in Fig. 5c and d. The boundary is found by finding the minimum argument for B. The boundaries then make it possible to compute the probability of wrong classification which is calculated as Eq. (2).

$$P_{wrong} = P(Water|Dry) + P(Ice|Dry) + P(Dry|Water) + P(Ice|Snow) + P(Snow|Ice) + P(Water|Ice) + P(Ice|Water) \tag{2}$$

For the wavelengths used in this investigation for the halogen light measurements P_{wrong} became 13.5%. The high wrong classification number is caused by the difficulty of separating the water and ice distributions.

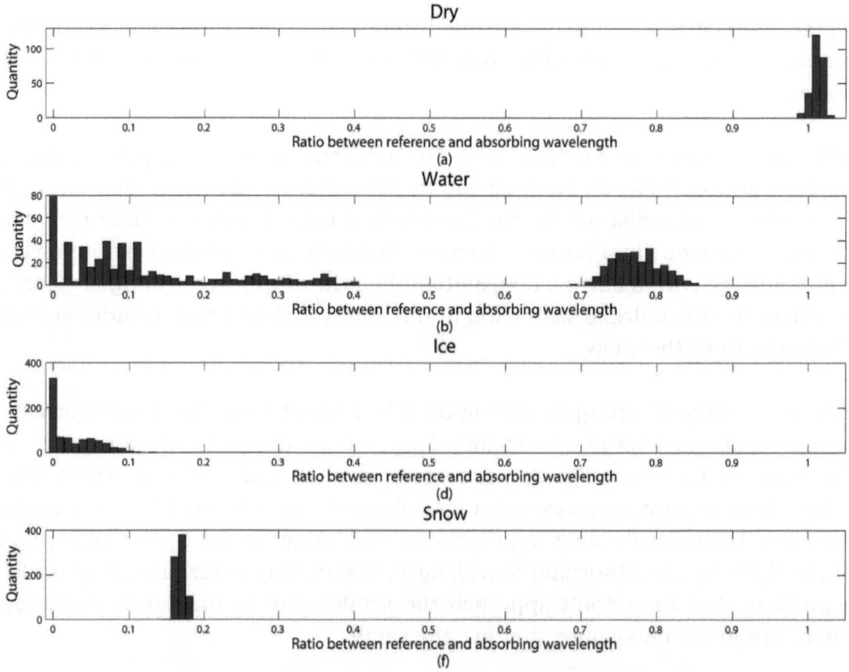

Fig. 7. The ratio distribution for the laboratory measurements with the Road eye sensor

3.3 Laboratory Measurements with the Sensor Road Eye

Notable in the result of the laboratory measurements with the Road eye as seen in Fig. 7, is that the width of the water distribution has increased compared with the first experiment with the halogen illumination. Note also that the ice distribution have decreased in width. This could be an effect of illumination with polarized and unpolarized light and different intensities of the illumination. Because of the polarization and the lower intensity in the laser-diodes, the Road eye is more sensible to variation in depths for the absorbing mediums water and ice. Separating the dry, ice and snow surface from each other is possible almost without any wrong classification. Though with the wide distribution of the water both the snow and ice distribution are effected which result in a high wrong classification

The water distribution have a wide range for a small change in the depths, this is the effect that causing the problems with the classification. The distribution that is located around 0.8 in the water histogram is the measurements for the moist asphalt, because of the difficulty of getting a continuous increase in the water depth it is a gap between the different measurements. The histogram for water should be continuous from 0.85 down to zero for the depth moist to 3 mm.

The wide distributions for both ice and water in the two laboratory experiments implies an absorbing medium as the depth increases the value decreases to zero. This is more distinct for the laser light which is polarized than for none polarized halogen illumination. The snow differs from the wide distributions of water and ice with a narrow distribution thou the depth were changed. This is an effect by the multiple scattering in the rough surface which is independent of the depth of the snow.

The probability of wrong classification calculation for the Road eye measurements results in a large uncertainty P_{wrong}=24%, this is due to the width of the water distribution. The dry and snowy asphalt is classified to 100%. The difficulty is separating water from ice, where the problem is when the depth increases both distributions approaches a zero value caused by the absorption of the light for the absorbing wavelength. Separating water and snow is also a problem, but they don't approach the same value as the depth increases. There are just a few values that are the same.

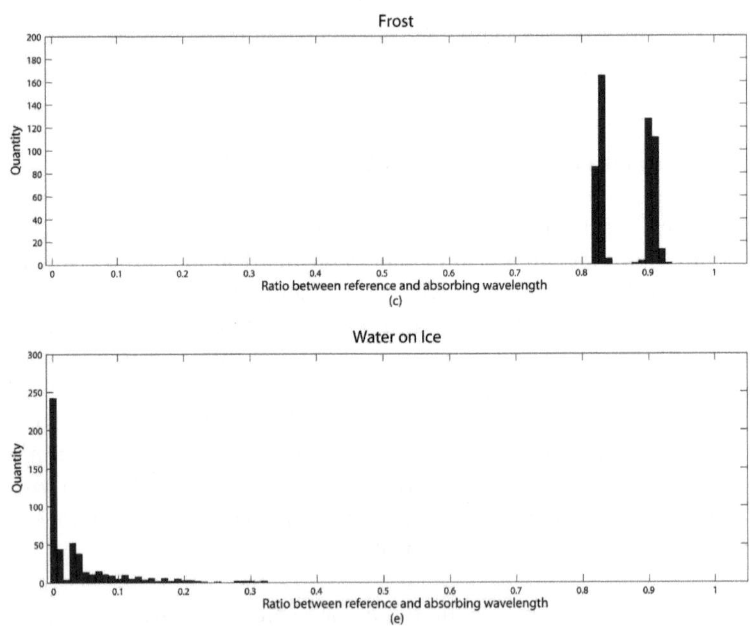

Fig. 8. The ratio distributions for the frost and water on ice measurements done with the Road eye sensor

In the laboratory measurements with the Road eye two more surfaces were measured. Seen in Fig. 8 is the result of the frost and water on ice. The two depths of frost were measured and it showed two spikes in the histogram, the gap between these two spikes implied that it should be a continuous distribution for a continuous change of depth. Which result in the problem that the frost could be classified as dry and wet asphalt as seen in Fig 8a) compared with Fig. 7a) and b). The high ratio for the frost also implies that the light is also reflected of the asphalt beneath the frost. The water and ice was classified as ice this due to the high depth of both water and ice which caused the ratio value to go towards zero (Fig. 8b).

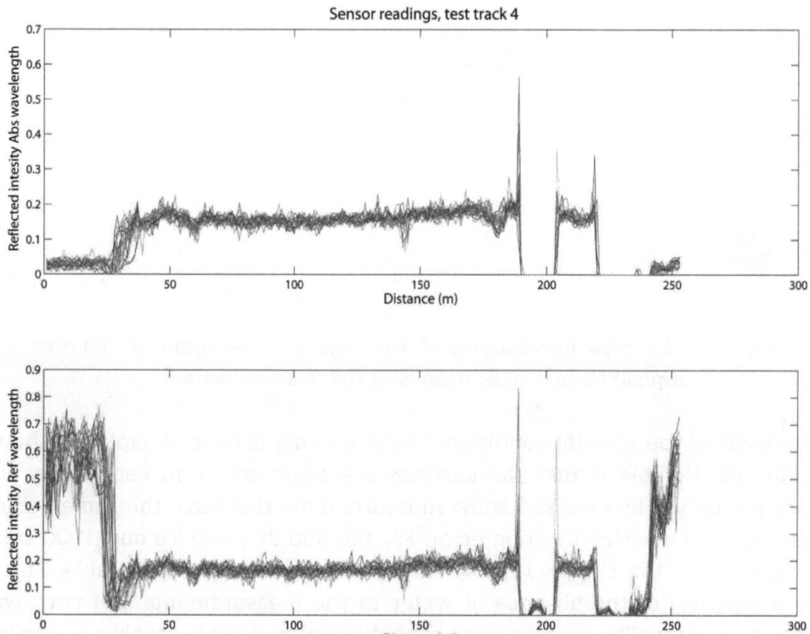

Fig. 9. Sensor readings of the test track 4 shown in Fig. 4 done with the Road eye sensor

3.4 Real-Life Condition Measurements with the Road Eye Sensor

The data for the two wavelengths computed for and plotted against the length of the test track is shown in Fig. 9. The result compared with the schematic figure in Fig. 4 implies a fast response time and a correct classification of the surfaces. This shows that the Road eye works acceptable in real-life condition measurements with defined surfaces.

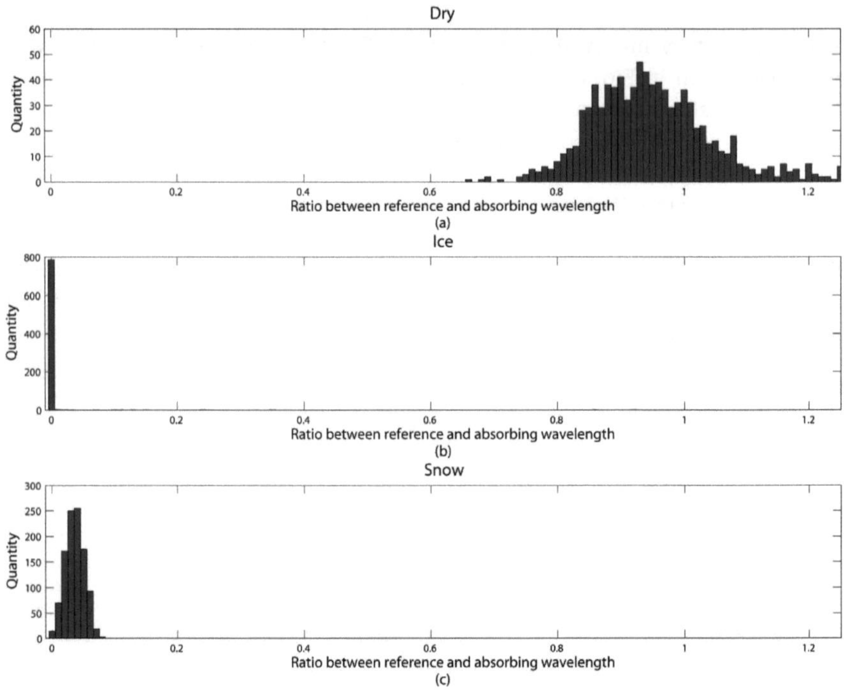

Fig. 10. The ratio distributions of the Arjeplog measurements for dry asphalt ice and snow done with the Road eye sensor

The result of the real-life condition measurements done in Arjeplog is shown in Fig. 10. Notable is that the surfaces are separated from each other and there are only a few ice and snow measurements that have the same values, this implies a low classification error. For the 900 dry, 800 ice and 1000 snow measurements the $P_{wrong} = 1.4\%$ which can be regarded as acceptable. This is a consequence of the absence of water in the measurements and very well defined surfaces. The ice was more than 40 mm thick which makes the reflection of the absorbing wavelength go to zero as seen in Fig. 9b, this explains the spike at zero for the ice.

The wide distribution of the asphalt seen in Fig. 10a may be caused by snow that was got caught on the trucks tiers and on to the dry asphalt and melted. But this amount of water on the asphalt doesn't affect the friction. The spike distribution of the snow could be caused by the compression of the truck and tiers that were hot and created a thin ice film on the snow. This also explains why the snow distribution is closer to zero for these measurements than for the laboratory experiments.

4 Conclusions

The conclusions that can be made of these measurements are that the three surfaces dry, ice and snow can be classified with a low probability of wrong classification. Because of the wide range of the water distribution this makes the classification much more uncertain, this problem can be solved using a third wavelength around 1700 nm as shown in [2] for laboratory measurements with the halogen light P_{wrong} decreases to around 1%. Another thing that separates water form ice and snow is the temperature on the surface so one possibility is to incorporate an IR-sensor that measures the surface temperature in the application. The real-life condition measurement shows that the Road eye works acceptable in working order with defined surfaces.

References

[1] Wallman C-G and Åström H: "Friction measurement methods and the correlation between road friction and traffic safety", VTI meddelande 911A, (2001)

[2] Casselgren J, Sjödahl M. and LeBlanc J.: "Angular spectral response -from covered asphalt" Applied Optics (2007)

J. Casselgren, M. Sjödahl
Division of Experimental Mechanics Luleå University of Technology
Luleå University of Technology
971 87 Luleå
Sweden
johan.casselgren@ltu.se
Mikael.sjodahl@ltu.se

S. Woxneryd, M. Sanfridsson
Volvo Technology Corporation Mechatronics and Software
Sweden
sara.woxneryd@volvo.com
martin.sanfridsson@volvo.com

Keywords: road condition, preview sensor, classification, asphalt, friction number

New European Approach for Intersection Safety – Results of the EC-Project INTERSAFE

K. Ch. Fuerstenberg, IBEO Automobile Sensor GmbH
J. Chen, St. Deutschle, Institut für Kraftfahrwesen Aachen

Abstract

The INTERSAFE project was created to generate an European approach to increase safety at intersections. A detailed accident analysis was carried out. Based on the derived relevant scenarios driver assistant functions are developed to support the driver in critical intersection situations. In addition evaluation and user test results of the Intersection Driver Warning System are presented and discussed.

1 Introduction

In the 6th Framework Program of the European Commission, the Integrated Project PReVENT includes Intersection Safety. The INTERSAFE project was created to generate a European Approach to increase the safety at intersections. The project started on the 1st of February 2004 and will end in January 2007.

The partners in the INTERSAFE project are:

- Vehicle manufacturer: BMW, VW, PSA, RENAULT
- Automotive supplier: TRW, IBEO
- Institute / SME: INRIA, ika/ FCS, Signalbau Huber

The main objective of the INTERSAFE project is to"....."

- improve safety and to reduce (in the long term avoid) fatal collisions at Intersections

In order to identify the most relevant scenarios for accident prevention, a detailed accident analysis was carried out. Based on the scenarios and the driver mistakes derived from the accident analysis a basic functionality is described. It considers for example the time budget, which is available in order to warn the driver.

The importance of these accidents leads to a deeper analysis of the scenarios. An in depth analysis of available data from reconstructed accidents in France and Germany show the central position of especially two accident types:

- collisions with oncoming traffic while turning left and
- collisions with crossing traffic while turning into or straight crossing an intersection.

Additionally the importance of the actual right of way regulation leads to the consideration of traffic light controlled intersections. Altogether, about 2/3 of all accidents in intersections are covered directly by the selection of these three scenarios. The possible coverage of other comparable accidents needs further investigations.

This paper will describe the INTERSAFE approach of intersection safety and discuss the functionality to avoid fatal accidents at intersections for the chosen scenarios in detail. Furthermore the evaluation results are presented.

2 INTERSAFE Concept & Vision

The INTERSAFE project realises two different approaches in parallel.

The first approach is a Bottom-Up Approach, based on state-of-the-art sensors and Vehicle-to-Infrastructure (V2I) communication, as shown in Fig. 1. Furthermore, some communication modules are installed at selected intersections in public traffic to realise the bidirectional communication between the vehicle and the traffic lights. This approach results in a basic intersection system, which can be evaluated in public traffic at selected intersections.

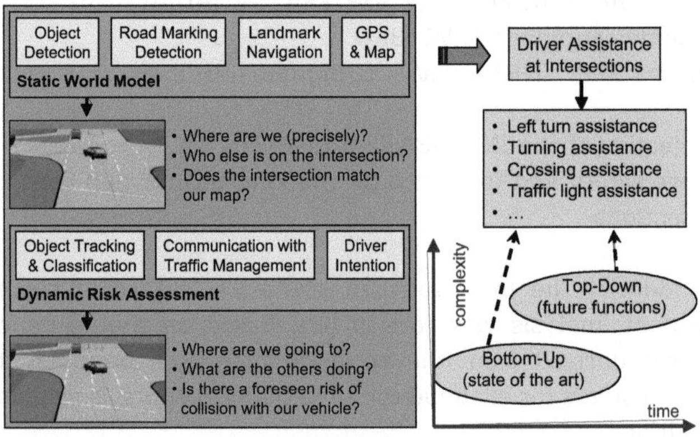

Fig. 1. INTERSAFE concept and vision

The second approach is a Top-Down Approach, based on a driving simulator. The driving simulator allows the analysis of dangerous situations, independent of any restricted capabilities of the sensors for environmental detection. The results of this approach are used to define an advanced intersection safety system, including requirements for future systems.

Fig. 2. BMW dynamic driving simulator.

Demonstrators

The INTERSAFE VW Phaeton Demonstrator is equipped with two laserscanners, one video camera and additional communication systems, as shown in Fig. 3. The video camera is used to process data about lane markings at the intersection while the Laserscanner collects data of natural landmarks as well as data about the road users. Highly accurate vehicle localisation is performed by fusion of the outputs of the video and laserscanner systems based on a detailed map of the intersection (high-level map).

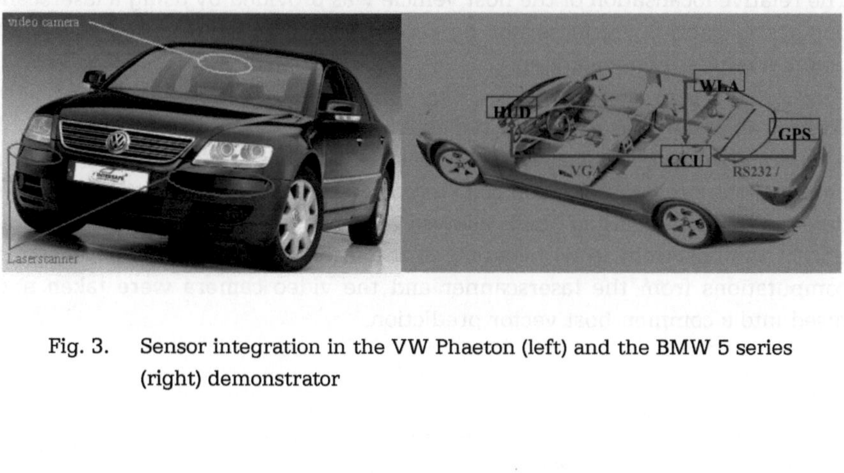

Fig. 3. Sensor integration in the VW Phaeton (left) and the BMW 5 series (right) demonstrator

The BMW demonstrator vehicle was based on a standard 5 series vehicle [5]. For its use as research prototype it was equipped with additional systems like V2I-communication or precise D-GPS-positioning. The main components are shown in the following Fig. 3. According to the HMI specifications information and warnings given in highly-complex intersection situations must be intuitive. Thus a head-up display was an appropriate way to provide information in the driver's field of view.

In a second step, a dynamic risk assessment is done. This is based on object tracking and classification, communication with the traffic management and the intention of the driver. As a result of the dynamic risk assessment, potential conflicts with other road users and the traffic management can be identified.

The laserscanner detects and tracks the road users which are essential for the functionality. Nevertheless, the video processing data and the fusion with the laserscanner data is crucial for the correct functioning of the host localisation process [2].

In the first section of this paper the localisation of the host vehicle and the Laserscanner based object detection, tracking and classification is described. This serves as the input level for the scenario interpretation and risk assessment which is described in the second part.

Localisation of the Host Vehicle

GPS-based localisation is generally not able to provide sufficiently accurate or reliable localisation in urban areas where intersections are typically located. The relative localisation of the host vehicle was provided by using a laserscanner system and a vision sensor by comparing current measurements with landmarks in a high-level map.

First of all, each sensor system (video camera and laserscanners) performed a separate relative localisation of the host vehicle based on its current sensor data and the previously offline registered road markings and landmarks. The detection and tracking of them allowed a calculation of position and orientation of the host vehicle with regard to the intersection. The two localisation computations from the laserscanner and the video camera were taken and fused into a common host vector prediction.

Object Detection, Tracking and Classification

In order to meet the functional requirements of monitoring crossing traffic two precise laserscanner are integrated into both left and right front corners of the demonstrator vehicle. Thus a combined scan area of 220° degree around the vehicle is achieved. Based on the laserscanner data detection, tracking and classification of road users are performed.

In the beginning the generated range profile is clustered into segments. Comparing the segment parameters of a scan with predicted parameters of known objects from the previous scan(s), established objects are recognised. Unrecognised segments are instantiated as new objects, initialised with default dynamic parameters.

In order to estimate the object state parameters a Kalman filter is well known in the literature and used in various functions, as an optimal linear estimator.

Object classification is based on object-outlines (static data) of typical road users, such as cars, trucks/buses, poles/trees, motorcycles/bicycles and pedestrians. Additionally the history of the object classification and the dynamics of the tracked object are used in order to support the classification performance.

Validation Methodology

Fig. 4 shows a general diagram of an intersection safety system and the internal system interfaces connecting each system level. Examples for transferred signals are shown. The figure simplifies the architecture and depicts on which system level the tests evaluate the system performance.

Starting with the technical verification, the output of the environmental sensors (laserscanner and video) and the communication module is monitored. Here the data are compared to the values described in the technical specification. At interface 1 (perception level) the output of the relative localisation of each sensor system, the detection, tracking and classification of road users and the V2I communication is evaluated.

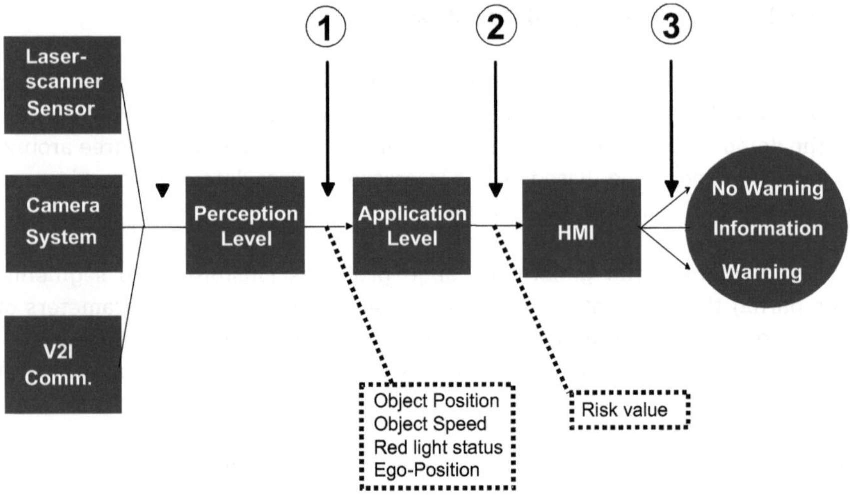

Fig. 4. General system description [3]

The second phase deals with operational verification. Here the system is inspected at interface 2 (application level). In addition, the implementation of the components is evaluated with respect to the operational specification.

Finally, the user aspects are addressed in the third part of the tests, the user test. The system is tested at interface 3 and tests with subjects are performed. The subjects performed test drives as well as the assessment of the system responses and the HMIs. Within this phase the same scenarios as in phase 2 are considered.

Validation Results

The main validation results for the demonstrator vehicles achieved in the INTERSAFE project will be described in the following sections [4].

Sensor Test

In the first evaluation phase, the sensors were tested to check the function of the INTERSAFE system at the perception level. In order to get a representative result, diverse objects were used as sensor targets in the tests. These targets were: VW Golf (silver estate car), VW Lupo (black compact car), BMW 325i (red middle size car), BMW 728i (black large size car), Honda VFR800 (silver motorcycle), pedestrian (dark clothing) and a wooden dummy target for the test of position accuracy.

Detection Range of the Laserscanner

In this test the laserscanner maximum detection range for all five test vehicles and the pedestrian was determined. The test started with the target vehicle moving towards the standing demonstrator vehicle.

Three different approaching directions of the target vehicles were applied in this test: frontal, 45° and 90° see (Fig. 5). If the opponent vehicle is perpendicular and far away from the host vehicle, it will typically leave the intersection before the host vehicle enters. Therefore the maximum detection range of perpendicular vehicles was just additional and performed with the VW Golf.

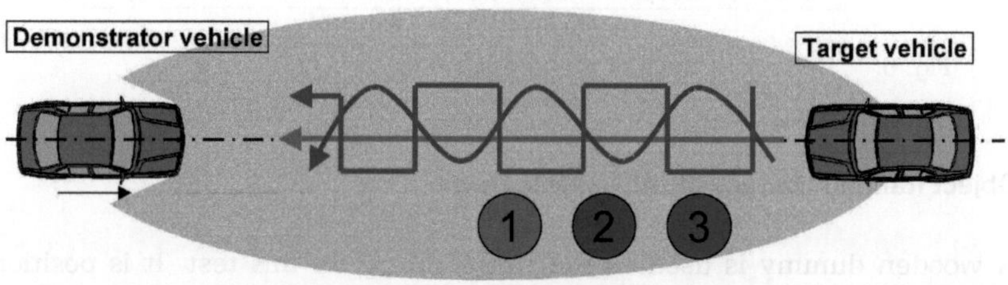

Fig. 5. Test layout

In order to avoid coincidences with regard to repeatability, every test was carried out twice for each target and approaching direction.

The maximum detection ranges of all the frontal and 45° tests are illustrated in Fig. 6. The CAN specification in the VW Phaeton demonstrator was limited to 200 m; a higher distance was not necessary. Therefore in this test, the maximum detection range of the laserscanner reported is 200 m.

During the tests, all the vehicles were correctly classified immediately after detection. All the cars were detected at a distance of more than 200 m, both in frontal and 45° tests. Just the maximum detection range of the motorcycle was slightly shorter.

The detection of pedestrian was also tested. Result showed that the pedestrian was detected and correctly classified as a pedestrian at distances of 110 m. In the 90° tests, the VW Golf was detected at about 165 m.

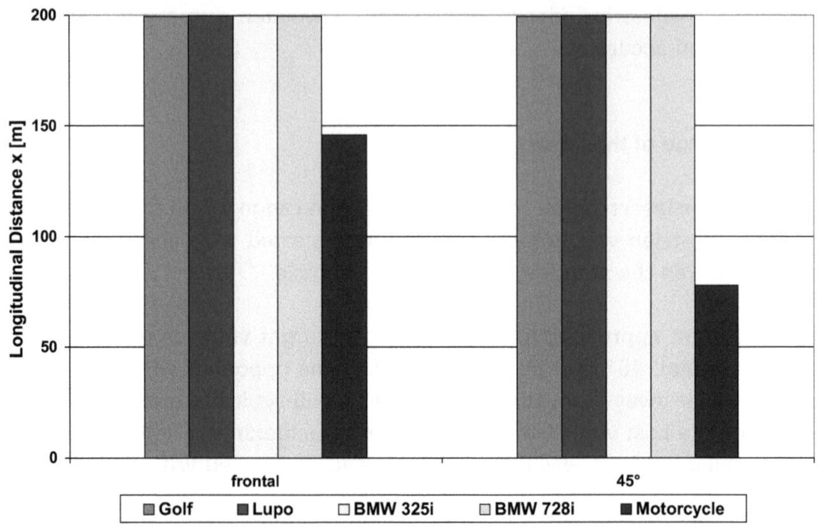

Fig. 6. Detection range of the laserscanner

Object Range Accuracy of the Laserscanner

A wooden dummy is used as the sensor target in this test. It is positioned either directly in front of the demonstrator vehicle or in a 45° position. In order to determine the accuracy at different distances, the demonstrator vehicle stands still and the target is moved in every test. The relative distance of the target is measured by the laserscanner and a reference device. The reference line is located around 15 m in front of the demonstrator vehicle's rear axle (the zero axis of the laserscanner output). The target is moved from this position until 50 m in five metre steps; namely 20 m, 25 m, 30 m, 35 m, 40 m, 45 m and 50 m.

The test results are illustrated in Fig. 7. The average error is about 10 cm.

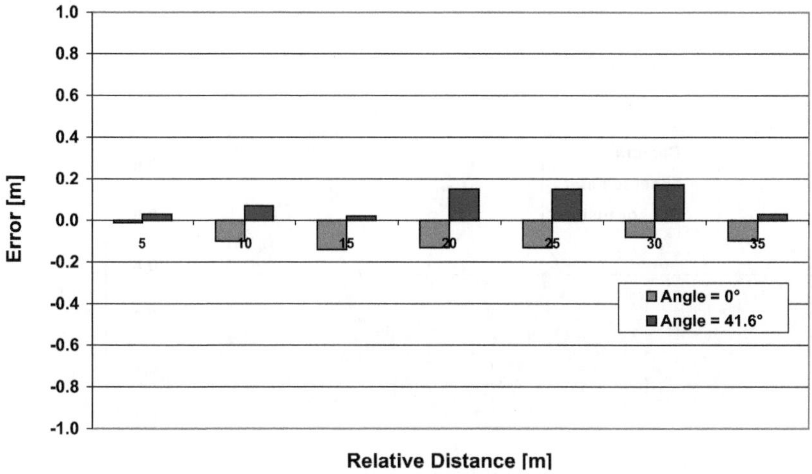

Fig. 7. Object range accuracy of the laserscanner

Localisation Accuracy

This test was applied to inspect the localisation accuracy of the laserscanner, the video system and the fusion output. The laserscanner system localises the vehicle position by detecting landmarks while the video system utilises road markings. As an example, the test for evaluation of longitudinal distance was described here.

In order to determine the demonstrator vehicle's distance to the intersection, a microwave sensor and a light barrier sensor were mounted at the rear end of the car. Four reflectors were put on the ground as reference positions for the light barrier. The distances to the intersection of these references were 10 m, 30 m, 50 m and 70 m respectively. According to the vehicle speed measured by the microwave sensor and the reference positions by the light barrier, the vehicle position could be calculated precisely.

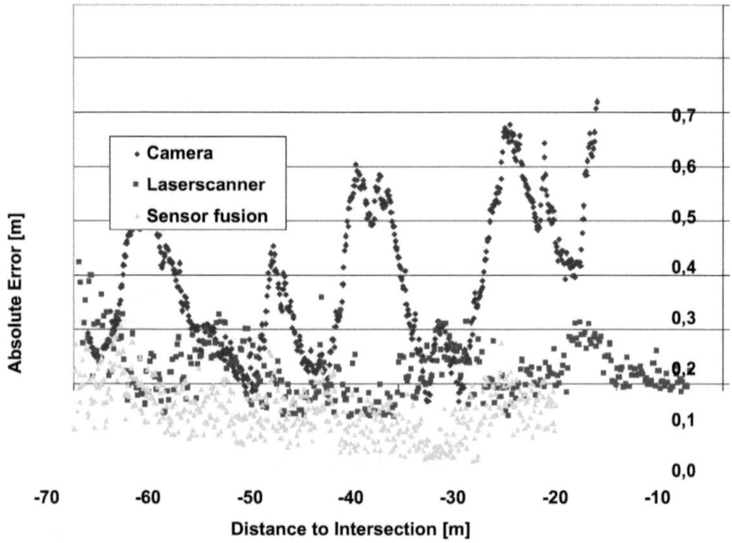

Fig. 8. Localisation error over distance for the camera (blue), laserscanner (red) and the fusion system (yellow)

The test was carried out five times. The localisation error for one exemplary test run is shown in Fig. 8. The average absolute errors of each test as well as the average of all the tests are summarised in Fig. 9. The outputs of all localisation systems were continuous, meaning no signal drop-outs occurred.

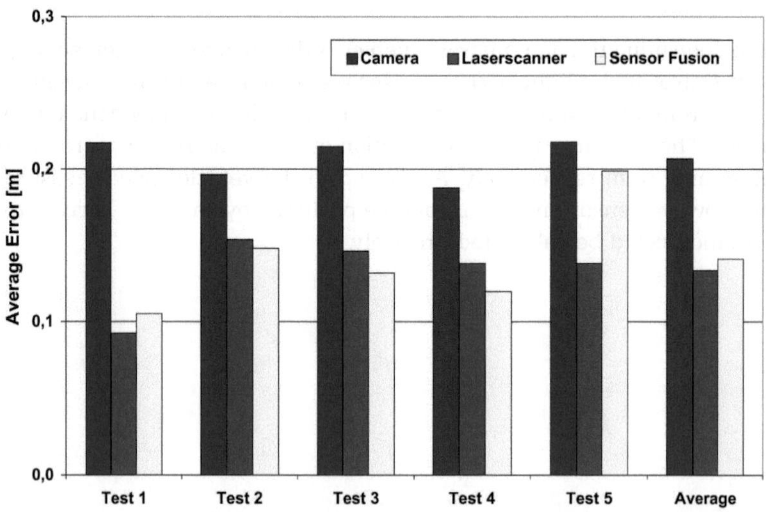

Fig. 9. Localisation accuracy distance for the camera (blue), laserscanner (red) and the fusion system (yellow).

System Test

The INTERSAFE system functions at the application level were tested in this evaluation phase. The evaluations were carried out based on the number and the rate of correct alarms, false alarms and missing alarms.

During the test, the demonstrator vehicle was driven through the intersection and the functions of the system were evaluated in the form of a check list.

The results of system test indicated that the Intersection Assistant had a correct alarm rate of 93% in left turn scenarios and 100% in lateral traffic scenarios. Both traffic light assistant systems achieved together an average correct alarm rate of 90%.

User Test

Sixteen subjects had been selected by taking their age, gender and driver experience into account. Each subject took around 2.5 hours to drive the demonstrator vehicles on ika's test track and assessed the performance of the INTERSAFE systems.

The assessment was realised by means of questionnaires. Subjects were asked to fill out a pre-questionnaire before the driving test, three questionnaires during the test and one post-questionnaire after the test.

The results regarding the helpfulness of the systems are shown in Fig. 10. The centre line means the average value while the lower and upper lines mean the average plus/minus standard deviation.

As illustrated in Fig. 10, the subjects rated the INTERSAFE systems helpful and relieving, stated especially by male and older subjects. Traffic Light Assistant was rated more helpful than Intersection Assistant. Further analysis showed that the subjects thought the Intersection Assistant for left turn was more useful than for lateral traffic. They judged that INTERSAFE could have helped them in their daily driving and it was agreed that it would improve the traffic safety.

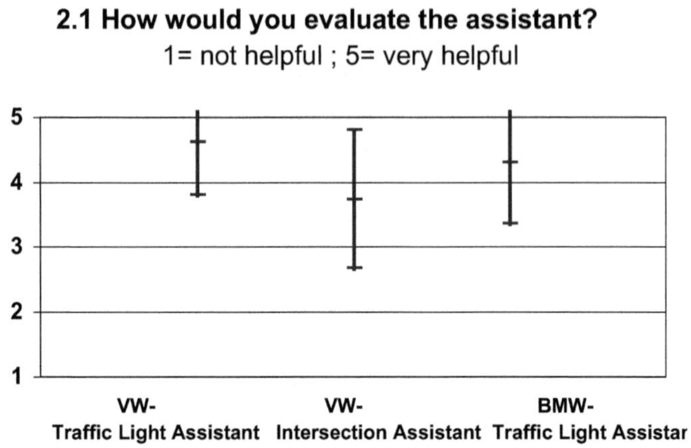

Fig. 10. Helpfulness of the intersection assistant systems

Conclusions

This paper describes the evaluation and testing of the two INTERSAFE demonstrator vehicles developed and implemented for intersection driver assistance. The tests were carried out in three evaluation phases: sensor test, system test and user test.

During the sensor test, sensors (laserscanner and video system) and the communication module were under investigation. The focus of this phase was to verify the sensor performance required for the INTERSAFE functions. The chosen sensors were fully suitable to fulfil the tasks in the INTERSAFE intersection driver assistance system.

The system test was carried out to check the system's functionality before the user test. Because of the system layout, a false-alarm-free result had not been achieved. In average, the correct alarm rate of the intersection assistant system was 97% and for the traffic light assistant 90% for both demonstrators.

In the last testing phase, sixteen subjects had driven the demonstrator vehicles on the test track and assessed the performance of the INTERSAFE intersection driver assistance systems. Generally (and in particular, male and older subjects) think the INTERSAFE systems are helpful and relieving. Subjects think traffic light assistant is more helpful than intersection assistant. Intersection assistant for left turn scenarios is rated more useful than the one for lateral traffic scenarios.

The INTERSAFE systems would have helped the subjects in their daily driving and it was agreed that it would improve traffic safety and that they want to have them in their car.

Acknowledgement

INTERSAFE is a subproject of the IP PReVENT. PReVENT is part of the 6th Framework Programme, funded by the European Commission. The partners of INTERSAFE thank the European Commission for supporting the work of this project.

References

[1] Fuerstenberg, K. Ch.: Intersection Safety – The EC-Project INTERSAFE. Proceedings of ITS 2005, 12th World Congress on Intelligent Transport Systems, ITS 2005 San Francisco, USA.
[2] Heenan, A.; Shooter, C.; Tucker, M.; Fuerstenberg, K. Ch.; Kluge, T.: Feature-Level Map Building and Object Recognition for Intersection Safety Applications. Proceedings of AMAA 2005, 9th International Conference on Advanced Microsystems for Automotive Applications, March 2005, Berlin, Germany.
[3] "INTERSAFE Validation plan", INTERSAFE Deliverable D40.77, Brussels, 2006.
[4] "INTERSAFE Evaluation results", INTERSAFE Deliverable D40.71/2/3/4, Brussels, 2007.
[5] "INTERSAFE Final Report", INTERSAFE Deliverable D40.75, Brussels, 2007.

Kay Ch. Fuerstenberg
IBEO Automobile Sensor GmbH
Fahrenkroen 125
22179 Hamburg
Germany
Kay.Fuerstenberg@ibeo-as.com

Jian Chen, Stefan Deutschle
Institut für Kraftfahrwesen Aachen
Steinbachstr. 7
52074 Aachen
Germany
Deutschle@ika.rwth-aachen.de

Exploiting Latest Developments in Signal Processing and Tracking for "Smart" Multi-Sensor Multi-Target ACC

M. Maehlisch, K. Dietmayer, University of Ulm
O. Loehlein, W. Ritter, DaimlerChrysler

Abstract

This contribution presents an overview of a whole detection and tracking system designed for supporting future smart ACC applications. The key system features are the usage of the latest signal processing and tracking algorithms addressing the major problems of today's available systems. The detection scheme based on sensor fusion of a multi-beam Lidar and a vision sensor is presented as well as a tracking algorithm fusing data from both sensors for refined state estimates and improved detection performance.

1 Introduction

The first generation of commercial ACC systems was introduced several years ago. These systems, either realized with a single Radar or a single Lidar sensor had been designed for highway use only. In order to detect relevant control targets, these systems evaluate features like absolute target velocity over ground and the track age. Major drawbacks resulting from this basic detection approach and the small surveillance area of the utilized sensors are the limited performance in high curvature environments and the inability to detect standing vehicles and vehicles on other lanes. Current research and development aims at building a situation aware ACC system that predicts expected manoeuvres and performs adequate reactions. Some of the addressed challenging traffic constellations are illustrated in Fig. 1. These future ACC systems will not only enable more economic control strategies with respect to fuel consumption and hybridization by avoiding unnecessary acceleration and braking periods, but also seriously improve customer acceptance.

To create such an application, some requirements need to be fulfilled by the underlying detection and tracking system. Obviously, the system must be real-time and multi-target capable in order to acquire a whole meta-state of several vehicles on different lanes. Moreover, the uncertainty of estimated dynamic states needs to be minimised in order to compute meaningful time forward predictions. A high detection performance in terms of false alarm quantity and detection rate is necessary, too, since i.e. heavy deceleration due to a false positive is a serious traffic incident with non-distinctive legal consequences. Because the basic detection strategy of accepting only targets with a significant estimated velocity and sufficient track age is inadequate in the questioned scenarios like traffic jams, the detection performance requirement implies robust sensor data with appearance features in bad weather and lighting conditions as well as the systematic incorporation of all available prior knowledge like dynamic state constraints as well as infrastructure state constraints that are available from digital maps for instance.

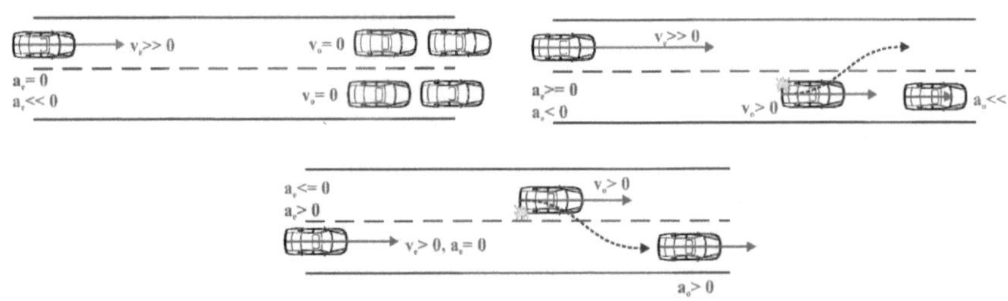

Fig. 1. Challenging traffic constellations demanding a situation aware ACC system; blue: current dynamic environment model, pink: expected manoeuvre, yellow: turn lights, red: first generation ACC reaction, green: "Smart" ACC behaviour.

The ability to recognize other road users in far distances is another required system feature, since early observation allows longer tracking periods and therefore tighter state enclosures on the first hand and enough time to react at high speed expressways on the other hand. Finally, the re-usage of sensors already available in serial production together with a low bandwidth communication network is an advantage related to packaging, integration and sales effort. This contribution presents a heterogeneous sensor fusion detection and tracking system designed to meet the specifications for supporting "smart" ACC applications as mentioned above.

1.1 Sensor Setup

The choice of fusing vision and ranging sensors below the detection level, the so called feature-level fusion approach, is very popular by reason of its numerous advantages. At first, appearance based detection can simultaneously rely on independent features from both sensors resulting in a noticeable enhancement of detection performance. Especially frequent ranging sensor false alarms emerging for instance from construction sites with crash barriers and reflectors or video false alarms, induced by background texture patterns with a certain similarity to vehicles can be easily eliminated in the combined feature space. Furthermore, the estimates of dynamic target state are expected to be more accurate and less uncertain, due to the synergetic combination of the excellent radial resolution of ranging sensors and horizontal and vertical angular resolution of vision sensors. This ensures stable and reliable velocity estimates with significant importance for state prediction. Moreover, a temporary dropout of a single sensor can be bridged by single sensor tracking with the other sensor without loosing the track. The addition of a vision sensor in contrast to single ranging sensor systems also allows the detection of turn signals, which further improves manoeuvre prediction.

The demonstration vehicle is therefore the equipped with an automotive CMOS camera as delivered with the serial nightview system. The camera is mounted behind the windshield, has a standard VGA resolution of 640 x 480, a maximum frame rate of 25 Hz, and is equipped with telephoto lens for magnifying distant objects. In addition, a multi-beam Lidar sensor is mounted into the radiator cowling. The Lidar emits 15 ns infrared Laser pulses at 905 nm in 16 fan-like arranged measurement channels. Each beam covers an azimuth angle of 1° and an elevation angle of 2.4°. The beam aperture azimuth is 1° too, resulting in a total Field of View (FOV) of 16°. The data update rate is fixed at 16 Hz and the detection range reaches up to 200 meters in favourable conditions. Further details about the Lidar are given in [1]. Fig. 2 shows the mounting positions of the sensors as well as their viewing cones.

Fig. 2. Test vehicle equipped with serial nightview camera and multibeam Lidar sensor.

The following sections deal with different subsystems of the detection and tracking system. At first strategies for sensor synchronization, a prerequisite of sensor fusion will be explained. The succeeding section presents the feature-level fusion vehicle detection process. Afterwards, the tracking stage estimating the host vehicle motion from ESP data and computing posterior estimates for both, state and existence of targets, is explained in section 4.

2 Sensor Synchronization

In order to fuse data from different sensors, a transformation to a global fusion coordinate system is necessary. These fusion coordinates are advantageously vehicle coordinates originating in the middle of the rear axis. The transformations between the fusion coordinate system and the local sensor coordinates, that are (i, j)-image coordinates for the vision sensor and polar (r, ϑ)-coordinates for the Lidar can be computed from intrinsic and extrinsic sensor parameters. While intrinsic camera parameters can be computed with state of the art calibration techniques, novel methods to estimate the intrinsic Lidar parameters as well as the extrinsic spatial alignments between the two sensor coordinate systems and the vehicle coordinate system have been developed. A key feature of these methods is the simplicity of the utilized calibration objects. In a first variant, any flat wall in any arbitrary position can serve as a calibration object.

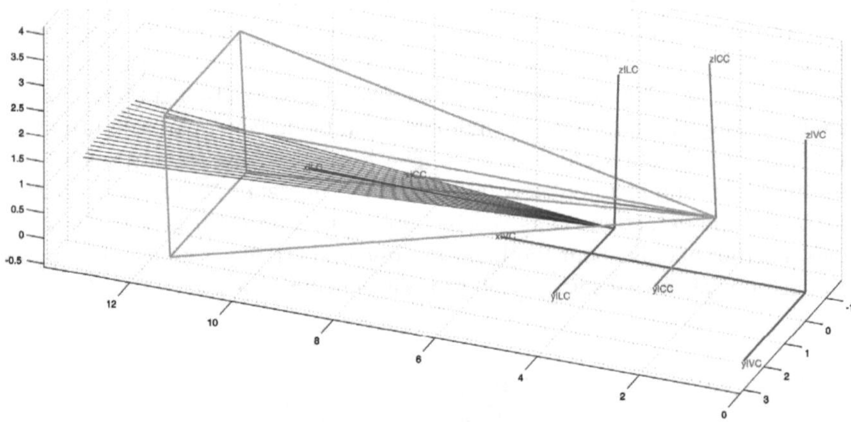

Fig. 3. Sensor models and spatial sensor alignment; pink: vehicle coordinate system (VC), green: camera coordinate system (CC) with viewing cone (cyan), red: Lidar coordinate system (LC) with medial beam axes (blue)

The basic idea of this very practicable variant is to capture the reflection spot positions of the Lidar beams in the image domain of the near infrared sensitive vision sensor. Afterwards, the position and/or orientation of the wall are changed relative to the vehicle and a second set of spot positions is recorded with the imager. After several iterations, it is possible to vary the alignment of the Lidar relative to the vision sensor, until the image projections of the Lidar beams (blue lines in Fig. 3) fit into the measured spot positions in the image domain. This method is described in detail in [2]. In a second variant, optical markers are attached to the calibration wall allowing to back-project the reflection spots into 3D camera coordinates. After collecting some sets of reflection spots with the vision sensor with different positions and/or orientations of the calibration wall it is now possible to fit the Lidar sensor model (blue lines in Fig. 3) directly into the 3D measurement points in local 3D camera coordinates. This allows the determination of intrinsic Lidar parameters like inter-beam alignments as well. Both calibration variants can be expressed as non-linear least squares fitters growing in accuracy with increasing number of measurement iterations. The temporal alignment is achieved by synchronising the camera exposure to the Lidar measurement times in hardware. Therefore both sensors operate with the Lidar measurement frequency of 16 Hz.

3 Vehicle Detection

The first vision based approaches to vehicle detection based on shape correlation with the edge image, symmetry assumptions and bumper shadow texture pattern matching. Although promising results were presented in the literature, real testing on the road revealed problems in bad lighting conditions, a lot of false positives in environments with complex background and missed detections in unfavorable conditions like infrastructure shadows from buildings and trees destroying the characteristic shadow patterns. Furthermore SUV's with spare wheels mounted at the rear side as well as imprints usually violate the symmetry assumption. This is why later systems used the mentioned "early" vision features for hypothesis generation and introduced a hypothesis verification stage based on pattern classification with neuronal networks, polynomial classifiers or support vector machines for instance.

The cascaded Adaboost classifier, introduced by Viola and Jones [3], systematically integrates the idea of early decision of most of the candidates with low effort and later verification of the few problematic candidates with increased computational cost. Besides this integrated "coarse-to-fine" strategy implied by the cascade architecture with growing complexity and cost from stage to stage, the Boosting strategy for choosing classification features in each stage classifier automatically selects relevant classifications features, models their statistical dependency and chooses alternative hints in case of weak primary features. Although the choice of features is suboptimal, due to the underlying Greedy strategy, it approximates the true separation function much better than any manually chosen decision rule set based on manually chosen classification features. This is why real on road testing shows cascaded Adaboost classification outperforming previous approaches.

In the field of single ranging sensor vehicle detection, sensors with poor angular resolution can only rely on the velocity constraint. Moderate angular resolution enables the classification by additional spatial echo amplitude and pulse-width pattern features and excellent angular resolution, as provided by Laser scanners, allows a classification by shape in near distances.

The first sensor fusion systems with video and ranging sensors followed a strategy of primary localization by a confirmed track from the ranging sensor and a succeeding state refinement with early vision features from the video sensor. Although exploiting the synergetic effects of sensor fusion for target state estimation, this approach does not improve the detection performance compared to single ranging sensor systems. Therefore the presented detection system uses the raw echo data from the ranging sensor as well as complex vision features based on pattern classification for decision making in a combined feature space.

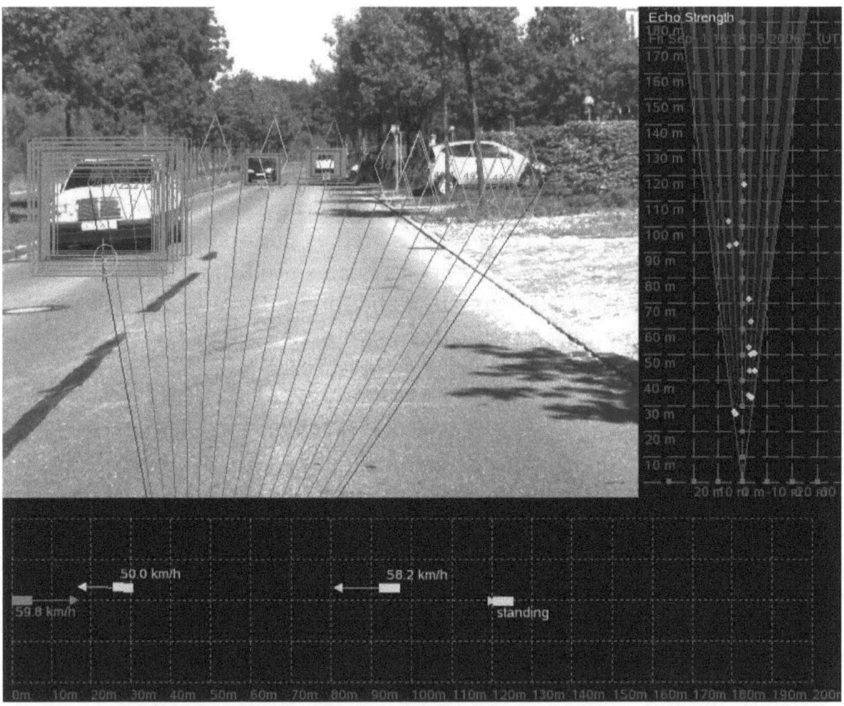

Fig. 4. Top-right: bird-view with Lidar echoes (yellow), top-left: video image with perspective Lidar beam (grey lines) and echo projections (diamonds), detection boxes of the cascade classifier (cyan boxes) and final detections (magenta boxes), bottom: environment model estimated by the tracking stage with host vehicle (green) and tracked vehicles (yellow)

The vehicle detection procedure starts by projecting the Lidar echoes (yellow dots in top-right bird view of Fig. 4) into the image domain. This can be done with the knowledge of the spatial sensor alignments determined with the presented synchronization procedures. The echo projections are illustrated as diamonds in the top-left video image of (Fig. 4). This step also requires the sensors to be temporally synchronized for the consistency of the image data and the attached depth information. Now the cascaded Adaboost sub window classifier, previously trained with a representative ground truth learn set, is applied to all image sub windows with a significant overlap with one of the projected Lidar echoes. As each Lidar echo provides a distance measurement, the image classifier hypothesis scaling problem can be solved without applying a flat-world-assumption. This attention control step enables real-time processing of standard VGA images.

The image classifier usually shows multiple detections for each true vehicle (cyan boxes in top-left video image of Fig. 4). Therefore a state-of-the-art clustering algorithm is used for merging. The final decision making is based on the image classifier outputs and the echo amplitude features provided by the Lidar. Two fusion strategies are suggested. A first method is to add the Lidar amplitude value to the feature pool and let the boosting classifier select and incorporate it automatically. Another variant is to explicitly insert another classification stage behind the image classifier cascade fusing the data from both sensors. We implemented the second method with a Bayesian classifier stage combining vehicle likelihood functions from both sensors. The Lidar vehicle likelihood function is approximated by the relative frequencies of the amplitude values for true and false positives of the ground truth database. The image vehicle likelihood function is determined in the same way, but a kind of "image amplitude" is derived either by the number of boxes in the detection clusters, the mean reached cascade layer in the detection clusters or the existence probabilities determined from a statistical model of the cascade classifier as described in [4]. If a sub window candidate passed the final detection stage, it enters the list of detected objects (magenta boxes in top-left video image of Fig. 4). An evaluation of this detection approach is given in [2].

4 Vehicle Tracking

Taking a view on a whole classical detection and tracking system architecture, consisting of a detection or segmentation stage, the data association stage, a succeeding tracking module and the final track validation, the primary task of the tracking stage is to filter the measured state variables, hence to reduce the sensor measurement noise, and to estimate temporal derivatives of state variables over time. From a mathematical point of view, the state uncertainty is reduced by combining prior knowledge from past measurements with the current sensor readings to a more precise posterior estimation. Besides the state uncertainty, which is treated in a Bayes-optimal way by recursive Bayesian state estimation like it is implemented with Kalman, Extended Kalman, Unscented Kalman and Particle Filters; there is another uncertainty domain in real applications with presence of clutter and missed detections. This uncertainty of existence or the uncertainty about the total number of targets in multi-target tracking is treated at least sub-optimal in the classical system architecture.

Most systems rely on a rule based track verification system for eliminating false positive tracks based on state constraints, filter consistency measures, the track age, the number of successful associations and other features. Other approaches append a trained classifier to the processing chain for decision making based on dynamic features estimated by the tracking stage in combination with the features of the detection stage. However, the existence uncertainty that becomes more important and harmful for the system output as the Signal to Noise/Clutter Ratio (SN/CR) decreases, is only handled in a heuristic way. Especially when using vision sensors, but generally in the task of appearance based classification, the uncertainty of existence becomes the dominating error source in complex environments. Most research in filtering and tracking focuses on sophisticated methods to determine the state of other road users more precisely while neglecting the basic problem of existence uncertainty.

A fundamental theory of joint multi-target modelling of state and existence uncertainty was introduced by R. Mahler and named Finite Set Statistics (FISST, [5]). Mahler's filtering equations allow the derivation of posterior estimates of multi-target state and existence but suffer from their inherent combinatorial complexity. Tractable algorithms approximating FISST theory based filtering are the Probability Hypothesis Density Filter (PHD, [6]) and the Integrated Probabilistic Data Association Filter (IPDA, [7]). All FISST derived filtering algorithms have in common, that besides the usual temporal evolution and measurement update models for target state they incorporate time forward prediction models and measurement update models for target existence as well and produce posterior Bayes-optimal estimates in both uncertainty domains simultaneously. Before dealing with the filtering equations, the next subsections shortly present the four necessary models for the ACC application.

4.1 Process Model for Target State

The state vector of other tracked vehicles was chosen as:

$$\underline{x} = (x, y, z, v, \psi, \dot{\psi})' \tag{1}$$

The position components (*x*, *y*, *z*) and the orientation angle Ψ are measured relative to the ego vehicle coordinate system, whereas the target velocity magnitude v and yaw rate *d*Ψ/*dt* are estimated absolutely over ground. The first step is to apply a constant turn movement model to the state vector describing the individual target motion (Fig. 5).

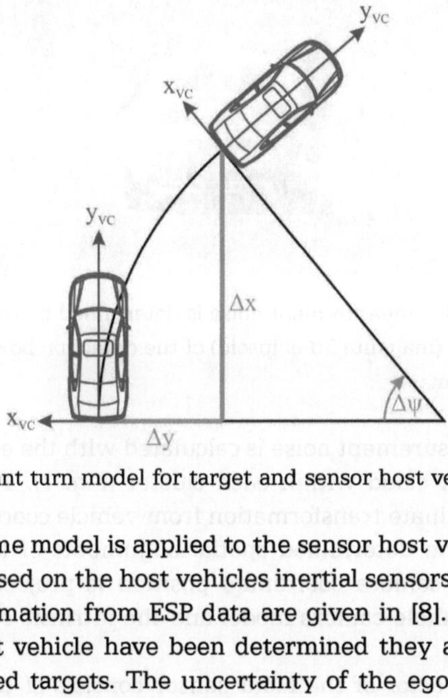

Fig. 5. Constant turn model for target and sensor host vehicle motion

Afterwards, the same model is applied to the sensor host vehicle using the ego motion estimate based on the host vehicles inertial sensors. Details on the host vehicle motion estimation from ESP data are given in [8]. After the shift and rotation of the host vehicle have been determined they are subtracted from the states of tracked targets. The uncertainty of the ego motion estimate is added to the state uncertainty of other targets. As a final step, the unknown host vehicle pitch angle adds uncertainty to the target states in the vertical direction (Fig. 6).

Fig. 6. State uncertainty induced by the vehicle pitching

For a detailed mathematical presentation of the state evolution model please refer to [9].

4.2 Measurement Model for Target State

The target state is updated from the Lidar and vision sensor in two independent steps. The vision measurement vector consists of the image coordinates of the baseline midpoints of the final detection boxes (Fig. 7, magenta cross).

Fig. 7. The video measurement noise is determined by the empirical covariance (magenta 3σ ellipsoid) of the detection boxes (cyan) within a cluster.

In addition, the measurement noise is calculated with the empirical covariance of the detection box cloud within each cluster (Fig. 7). The sensor model is composed of a coordinate transformation from vehicle coordinates into camera coordinates using the determined spatial alignments of the synchronization step. Next, the 3D camera coordinate position is projected into the image domain with the pinhole camera model and the intrinsic camera parameters.

The measurement vector of the Lidar sensor consists of the radial distance r and the azimuth angle of the measurement channel ϑ (Fig. 4, top-right). The measurement function h, mapping from state space to the Lidar measurement space is composed of a coordinate transformation from vehicle coordinates into Cartesian Lidar coordinates followed by a transformation from Cartesian Lidar coordinates into polar Lidar coordinates. The measurement noise uncertainties are assumed independent in polar Lidar coordinates and initialized with the radial position resolution specification from the sensor manufacturer and the angular resolution of the beam fan as standard deviations.

4.3 Process Model for Target Existence

The process model for target existence must provide the probabilities of existence state changes between two subsequent time steps, which are completely determined by the probability of target persistence and the probability of target birth.

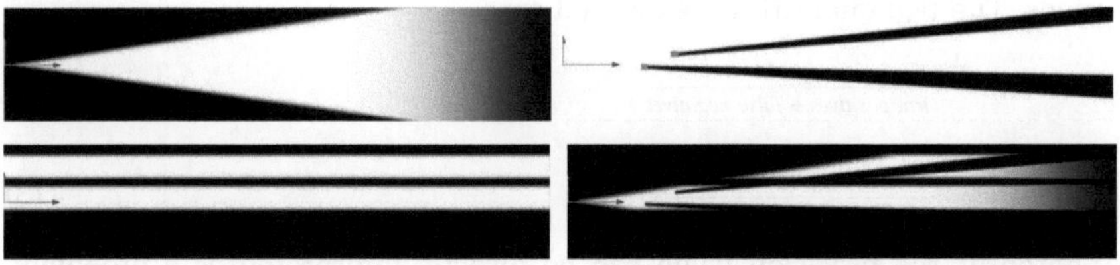

Fig. 8. Cardinal persistence model; top-left: state restrictions due to the sensor Field of View (FOV), top-right: surveillance shadows of confirmed targets, bottom-left: state restrictions from a digital map, bottom-right: combined persistence probability map, the probability interval 0..1 is colour-mapped from black to white.

The persistence probability is modelled dependent on the current target state and accounts for the sensor field of view, the mutual occlusion of vehicles and prior position constraints obtained for instance from digital maps. The top-left part of Fig. 8 shows the object persistence probability given by the Lidar field of view. The persistence probability inside the FOV is obviously one, due to the fact that targets cannot disappear spontaneously. The persistence probability induced by the surveillance shadows of confirmed targets is illustrated in the top-right part of Fig. 8. Finally a precise localization on digital maps as presented in [10] can introduce prior constraints on target position, as visualized in the bottom-left part of Fig. 8 for a two lane road piece. The position restrictions, allowing vehicles to exist only on roads, can also be obtained from optical lane recognition systems. Finally any other prior state constraints like maximum velocities can be incorporated into the existence prediction model. The final probability map of target persistence, given the current state is a superposition of all individual aspects as shown in the bottom-right part of Fig. 8. Since target birth is only possible at the boarders of the FOV and surveillance shadows, the birth probability map can be modelled proportional to the spatial gradient vector magnitude of the combination of the FOV and occlusion persistence probability maps. Details of the existence process modelling can be found in [11].

4.4 Measurement Model for Target Existence

The measurement update requires a probability measure of target existence from the detection algorithm as well as the probability of detection. The easiest way to obtain these quantities is the analysis of the Receiver Operating Characteristics (ROC) of the detector. Usually ROC curves are generated by recording sensitivity and precision values for different algorithm parameterizations. The two quantities are defined as:

$$sensitivity = \frac{true\,positives}{true\,positives + false\ negatives} \tag{2}$$

$$precision = \frac{true\,positives}{true\,positives + false\ positives} \tag{3}$$

Apparently, the precision value can be directly interpreted as a probability measure of target existence and the sensitivity is the probability of detection. Therefore, all required measures can be extracted from the current detector operating point in the ROC chart. This is a generic interface, usable with any detection strategy. The ROC chart of the detection scheme presented in section 3 is shown in Fig. 9.

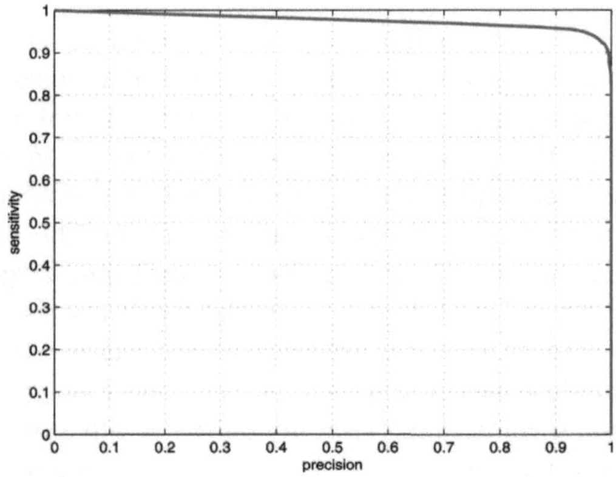

Fig. 9. Receiver operating characteristics chart for the detection scheme presented in Section 3.

Current research goes beyond these common probability measures. So called probabilistic detectors provide measurement–individual existence probabilities, which are expected to further improve the system performance. A promising approach for the cascaded Adaboost detector utilized in our detection scheme is the statistical modelling of the algorithm as a probabilistic Boosting tree. Possibilities to extract true existence probabilities per detection have been examined in our research group [4] and will be integrated in the existence measurement update model of FISST based tracking in the future.

5.5 Filtering Algorithms

As a reference system, the classical system architecture using multi instance Extended Kalman Filtering has been implemented. A data flow graph of the whole detection and tracking system is given in Fig. 10.

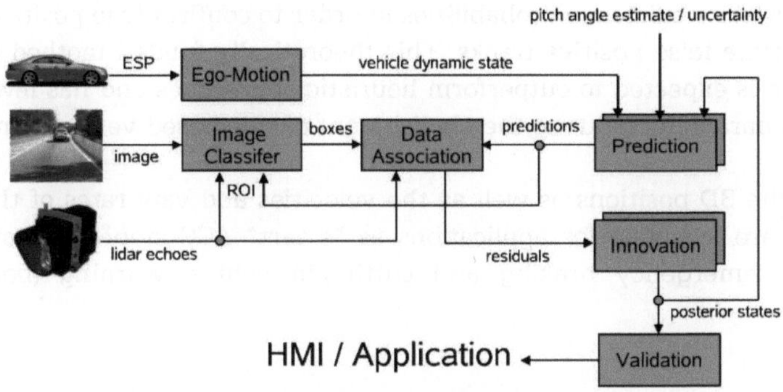

Fig. 10. Data flow graph of the multi instance tracking variant as implemented by the IPDA-filter algorithm

A detailed discussion of the PHD filtering algorithm applied to automotive vehicle tracking can be found in [12]. In the following the IPDA filtering equations will be presented. Integrated probabilistic data association is a multi instance filtering approach that requires explicit data association. It models the temporal evolution of state and existence uncertainty as two separate cross-coupled Markov chains. The Markov chain for the state uncertainty is the ordinary Kalman filter and will not be presented. The IPDA existence time forward prediction formula is given by:

$$p_{k|k-1}(e) = p_p(x_{k|k-1}) \cdot p_{k-1|k-1}(e) + p_b(x_{k|k-1}) \cdot (1 - p_{k-1|k-1}(e)) \tag{4}$$

Here, $p(e)$ is the probability of target existence, x is the current target state prediction, p_p is the probability of target persistence and p_b is the probability of target birth. The measurement update equation modifies the existence probability according to the number of associated measurements m_k:

$$p_{k|k}(e) = \frac{(1-\delta)\, p_{k|k-1}(e)}{1 - \delta \cdot p_{k|k-1}(e)} \tag{5}$$

$$\delta = p_d - p_d \frac{V_k}{\lambda_V} \sum_{i=1}^{m_k} p(z_i \mid x) \tag{6}$$

$$\lambda_V = \max(0, m_k - p_d \cdot p_{k|k-1}(e)) \quad (7)$$

The probability p_d is the probability of detection taken from the ROC chart as explained before and $p(z|x)$ is the measurement likelihood function as used in the state estimation Markov chain. After filtering, a threshold can be applied to the posterior existence probabilities in order to confirm true positive tracks and to prune false positive tracks. This theoretically funded method of track validation is expected to outperform heuristic approaches and has fewer optimization parameters, due to the abolishment of rule based verification.

Finally, the 3D positions as well as the velocities and yaw rates of the other vehicles are available for applications as "smart" ACC stop&go, traffic jam warning, emergency braking and cutting-in vehicle warning (bottom of Fig. 4).

5 Conclusions

A detection and tracking system addressing the major drawbacks of today's ACC systems by incorporating latest signal processing and tracking algorithms has been presented. The system is implemented in a demonstration vehicle with serial night vision camera and a multi beam Lidar sensor. Even standing vehicles are detected in distances up to 150 meters and the system can keep track of already detected vehicles up to 200 meters. The current development phase allows non-synthetic in-vehicle demonstrations in real traffic scenarios.

5.1 Future work

Future development will evaluate the different FISST based tracking methods compared to the standard approach in terms of detection performance and state estimation accuracy. Another aim is to port the presented detection and tracking strategy to Radar-video sensor fusion and to other applications like night vision and pedestrian recognition.

5.2 Acknowledgement

This research was supported by NIRWARN (Near-Infrared Warning), BMBF 01M3157B, Germany.

References

[1] J. Thiem, and M. Muehlenberg: "Datafusion of Two Driver Assistance System Sensors", In Advanced Microsystems for Automotive Applications, 2005

[2] M. Maehlisch, R. Schweiger, W. Ritter, and K. C. J. Dietmayer, "Sensorfusion using spatio-temporal aligned video and lidar for improved vehicle detection," in Proceedings of IEEE Intelligent Vehicles Symposium, Tokyo, Japan, 2006.

[3] P. Viola and M. Jones, "Robust real-time object detection," in Second international Workshop on statistical and computational Theories of Vision-Modeling, Learning, Computing, and Sampling, Vancouver, Canada, July 13 2001.

[4] R.Schweiger, H.Hamer, and O. Loehlein, "Determining posterior probabilities on the basis of cascaded classifiers as used in pedetrian detection systems," Submitted to IEEE Intelligent Vehicles Symposium, 2007.

[5] R.P.Mahler, "Random sets: Unification and computation for information fusion - a retrospective assessment," in Proceedings of the Seventh International Conference on Information Fusion, P. Svensson and J. Schubert, Eds., vol. I. Mountain View, CA: International Society of Information Fusion, Jun 2004, pp. 1–20.

[6] R. P. Mahler, "Multitarget bayes filtering via first-order multitarget moments," in IEEE Trans. AES, vol. 39, no. 4, 2001, pp. 1152–1178.

[7] D.Musicki, R.Evans, and S.Stankovic, "Integrated probabilistic data association," IEEE Trans. Automatic Control, vol. 39, pp. 1273– 241, 1994.

[8] M. Maehlisch, W. Ritter, and K. Dietmayer, "ACC vehicle tracking with joint multisensor multitarget filtering of state and existence," Prevent ProFusion eJournal, vol. 1, p. 37ff, 2006.

[9] M. Maehlisch, W. Ritter, and K. Dietmayer, "Feature level video and lidar sensorfusion for full speed ACC," Submitted to 4th International Workshop on Intelligent Transportation, Hamburg, Germany, 2007.

[10] S. Wender, T. Weiss, and K. Dietmayer: "Object Classification exploiting High Level Maps of Intersections", In Advanced Microsystems for Automotive Applications, 2006

[11] M. Maehlisch, W. Ritter, and K. Dietmayer, "Decluttering with Integrated Probabilistic Data Association for Multisensor Multitarget ACC Vehicle Tracking," Submitted to IEEE Intelligent Vehicles Symposium, 2007.

[12] Mirko Maehlisch, Roland Schweiger, Werner Ritter, Klaus Dietmayer: "Multisensor Vehicle Tracking with the Probability Hypothesis Density Filter" Proc. of ISIF/IEEE 9'th International Conference on Information Fusion, Florence, Italy, 2006

Mirko Maehlisch, Klaus Dietmayer
University of Ulm, Inst. Of Measurement, Control, and Microtechnology
Albert-Einstein-Allee 41
89081 Ulm
Germany
mirko.maehlisch@uni-ulm.de
klaus.dietmayer@uni-ulm.de

Otto Loehlein, Werner Ritter
DaimlerChrysler Research and Technology, Dept. GR/EAP
Wilhelm-Runge-Straße 19
89081 Ulm
Germany
otto.loehlein@daimlerchrysler.com
werner.r.ritter@daimlerchrysler.com

Keywords: sensor fusion, cascade classifier, integrated probabilistic data association, smart ACC, finite set statistics

Enhancing ACC Stop&Go with Digital Map Information

W. Justus, R. Schulz, M. Köhler, Ibeo Automobile Sensor GmbH

Abstract

Lidar ACC Stop&Go is an advanced driver assistance system that was recently introduced. The system is capable to handle a large set of driving situations.

In this paper an extension to this driver assistance system using information from a digital map database is presented. The system behavior is improved and features are added by using an electronic horizon from the map data. Digital maps extend the range of the laserscanner with abstract information.

1 Introduction

Forward looking sensors offer a detailed representation of the environment. However, they are limited in sensing range. The laserscanner has a detection range of approx. 200 m, a wide horizontal field of view and a high angular resolution. Fig. 1 shows the sensing range and the field of view of a single Ibeo laserscanner. Digital maps deliver off-line data with quasi unlimited range. This information is more abstract and less detailed than the laserscanner data. Thus, digital maps providing additional information complement the description of the surrounding.Ibeo extends its ACC Stop&Go in cooperation with NAVTEQ with map support to optimize the system performance.

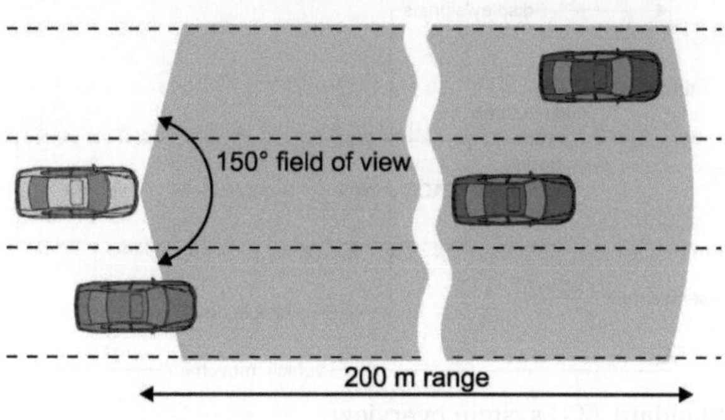

Fig. 1. Sensing range of Ibeo ALASCA XT

2 Lidar ACC Stop&Go

Standard ACC is a well known driver assistance system that eases driving by longitudinal control of a vehicle. ACC Stop&Go is capable of handling a largely increased set of driving situations on highways, rural roads and partly in urban environment [3].

However, driving scenarios exist that are not fully supported by ACC Stop&Go as the system is limited in e.g. road type classification and curvature prediction.

Fig. 2 shows the ACC system structure. The ACC controller uses laserscanner data, vehicle movement information like speed, yaw rate, steering angle and the input from a standard HMI.

Interpretation of the laserscanner data, vehicle data and parameters from driver interaction are available for the ACC system. All decisions during vehicle control have to be made based on that information.

The laserscanner system delivers a set of objects that are prioritized by the ACC controller. If available, one is selected as lead vehicle for the control algorithms. Determining the suited reaction on obstacles demands an adequate description of the surrounding.

The system behavior has to be comfortable, safe and adapted to the driving situation and the surrounding. Both are determined from the laserscanner data. Additional information from digital maps can help to assure and to extend that data. As drivers do not accept unforeseen braking maneuvers but expect prompt reactions if necessary, such supporting information is useful.

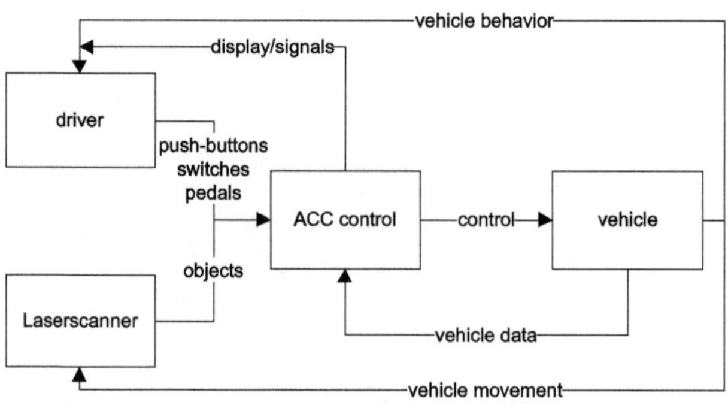

Fig. 2. Standard ACC system overview

3 Electronic Horizon from Digital Map Data

NAVTEQ's map database contains vector data with several attributes of a certain area like Europe. The grade of details depends on resolution respectively point density. Waypoints of roads have attributes like radius, road type, number of lanes and speed limit.

First the measured GPS position is matched to the road network, if driving on a known road. For refining tracking of the ego position - especially during missing GPS signal - external sensor information (vehicle sensors and extra gyro) is used.

The software in NAVTEQ's ADAS development tool calculates an electronic horizon providing information about the road ahead. Fig. 3 shows exemplarily the electronic horizon 2 km ahead of the ego vehicle on a German highway and the corresponding map. The ego vehicle is shown in yellow on the left of the electronic horizon and at its position on the map. The electronic horizon indicates the number of lanes, the exit and access road ahead.

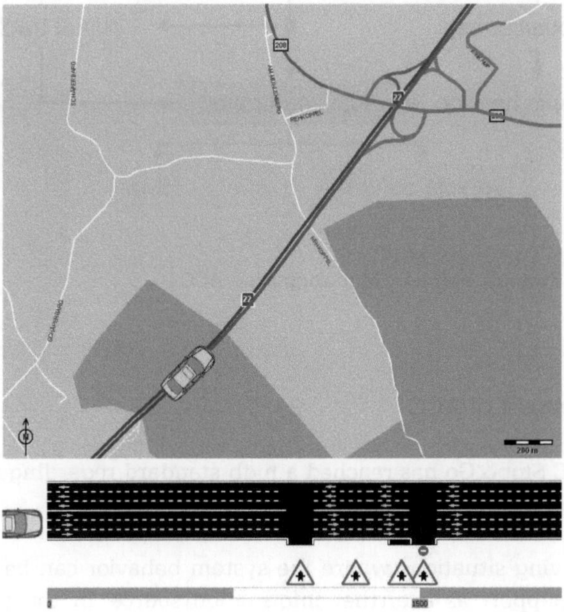

Fig. 3. Visualization of digital map data and electronic horizon from NAVTEQ

4 System Overview

Fig. 4 shows a schematic of the map supported ACC system. There are two separate systems processing the sensor inputs. GPS and vehicle data are used by the NAVTEQ software. The left part of the schematic shows how the electronic horizon is built using ego position and map database. Laserscanner and vehicle data are used by the Ibeo software to detect and to track objects.

The ACC controller receives information from both parts. Hence the dataflow is one-way from the different systems to the map supported ACC Stop&Go.

As map support is an addition for ACC Stop&Go, the system is also working without map data as fallback solution.

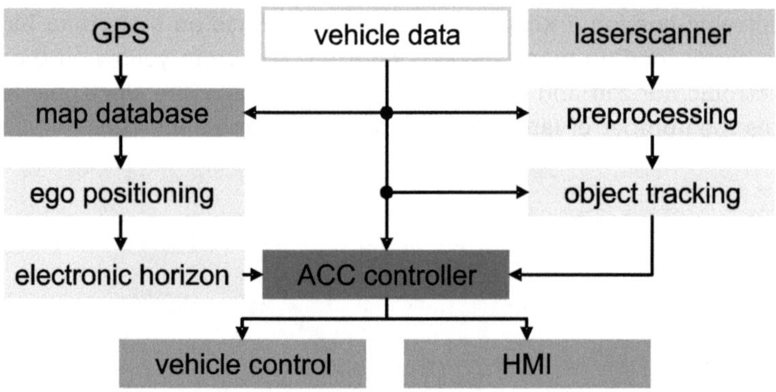

Fig. 4. System overview: Map supported ACC

5 Improvement of ACC

However, ACC Stop&Go has reached a high standard regarding comfort, supported driving situations, reliability and safety.

There are driving situations where the system behavior can be improved by adding map support as a further information source. In the following it is shown how ACC profits from digital map data support.

5.1 Road Curvature

With information about the oncoming road curvature, the selection of the lead vehicle becomes more accurate. E.g. approaching a curve that lies beyond the laserscanner range could lead to false decisions in the object selection algorithms. The predicted driving path of the ego vehicle (blue points) is straight as the ego vehicle hasn't entered the curve yet (Fig. 5). The red car is located in this path and hence is estimated as relevant. As well the lateral movement of the red car could be interpreted as cut-in scenario. In such a situation unexpected braking maneuvers happen frequently during ACC usage.

If the ACC controller has the information about an oncoming curve it expects lateral movement of the vehicles ahead when they enter the curve. Therefore the selection of the lead vehicle becomes more accurate. The driving path prediction is also supported by the road curvature.

Another feature is to reduce the set speed to a safe level before entering a curve. A safe cross acceleration is used to calculate the speed limit. Alternatively a warning can be issued to the driver instead of adjusting the set speed.

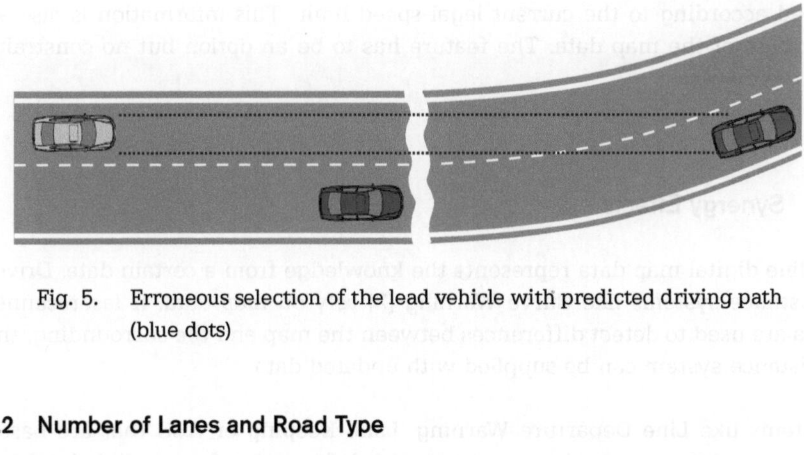

Fig. 5. Erroneous selection of the lead vehicle with predicted driving path (blue dots)

5.2 Number of Lanes and Road Type

If there is only one lane for the current driving direction, the selection of the lead vehicle can often be done easily. The closest vehicle ahead moving in the same direction as the ego vehicle is determined as lead vehicle. The driving path prediction becomes less important.

The current road type (e.g. highway, rural road or inner-city) and the number of lanes from the map data allows to enhance lane assignment of vehicles and obstacles. The ACC dynamics on different road types are adapted to the current driving situation with dynamic parameters [1].

5.3 Highway Exits

A standard problem of ACC systems is that they have no knowledge about the driver's intention. If the driver wants to leave a highway and enters the exit lane, frequently the lead vehicle is lost for the ACC controller due to the lane change. If there is no other vehicle that leaves the highway ahead, the decision of the system is to accelerate to the set speed that usually is inadequate for this situation.

If the ACC controller is informed that an exit is near and if a lane change to the right is detected, the speed is limited to a safe level.

5.4 Legal Speed Limit

Another additional feature is to provide an automatic adaptation of the set speed according to the current legal speed limit. This information is also an attribute of the map data. The feature has to be an option but no constraint for the driver.

6 Synergy Effects

Offline digital map data represents the knowledge from a certain date. Driver assistance systems like Curve Warning [2] rely on map data. If laserscanner data are used to detect differences between the map and the surrounding, the assistance system can be supplied with updated data.

Systems like Line Departure Warning, Lane Keeping or ACC that are based on sensor data can also be supported with information from a digital map to detect errors in sensor data interpretation.

It is also possible to cross-check map data with sensor information and to build an updated dynamic map around the vehicle.

Fig. 6 shows a video picture and laserscanner data of a scenario with a blocked lane. In the bird's eye view the ego vehicle is at the bottom (origin of the coordinate system). Driving direction is upwards. The three important objects (truck, car and barrier) are added to the electronic horizon below. The map is temporarily extended.

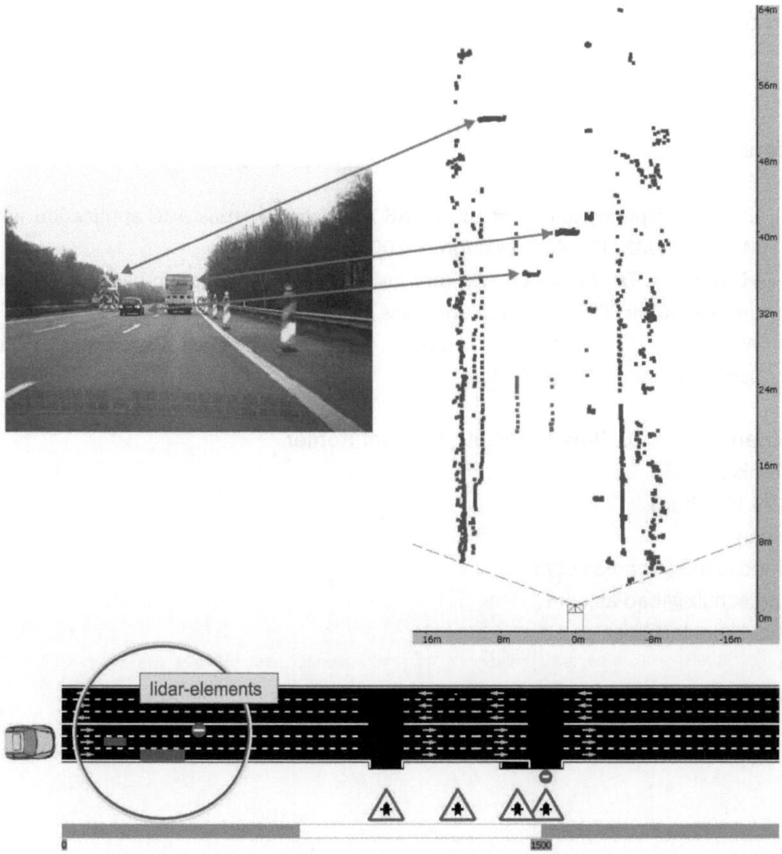

Fig. 6. Video picture of blocked lane, laserscanner data and updated electronic horizon

7 Conclusions and Outlook

Supporting ACC by digital map data is an effective way to enhance the system performance.

The described features help reducing wrong decisions of the ACC controller and hence the number of driver interventions during ACC usage. Therefore comfort is improved and ACC can be used in more driving situations.

In the currently used system the software in NAVTEQ's ADAS development tool is used without route planning. It generates the electronic horizon depending on the most probable driving path. As a next step routing will be used to foresee driving maneuvers.

References

[1] L. Beuk: Implementation of the ADAS Interface in series ACC application within MAPS&ADAS. ITS World Congress 2005, San Francisco

[2] M. Garrelts, Dr. M. Mittaz, A. Varchmin, W. Vogt: Navigation-based driver assistance systems. ITS European Congress 2005, Hannover

[3] W. Justus, V. Willhoeft, H. Salow, J. Kibbel: Full Speed Range ACC using a Laserscanner. WIT 2006, Hamburg

Winfried Justus, Dr. Roland Schulz, Michael Köhler
Fahrenkrön 125
22179 Hamburg
Germany
winfried.justus@ibeo-as.com
roland.schulz@ibeo-as.com
michael.koehler@ibeo-as.com

Keywords: ACC Stop&Go, ADAS, digital map support, laserscanner

Landmark Navigation for Robust Object Tracking in Skidding Maneuvers Using Laser Scanners

T. Weiss, St. Wender, K. Dietmayer, University of Ulm

Abstract

Many advanced driver assistant systems depend on a correct determination of the dynamic states of the host vehicle and of moving objects detected by environmental sensors. In order to estimate the correct position and orientation of tracked objects over ground, the movement of the host vehicle must be known precisely. The translation and the change of the yaw angle of the host vehicle during a short period of time are determined using integrated standard sensors, such as yaw rate, steering angle and wheel speed sensors. However, in extreme situations, such as skidding or wheel spin, the integrated sensors are not able to measure the host vehicle's real translation and rotation. In order to provide a valid position and orientation information even in these extreme situations especially for object tracking systems, stationary objects in the environment of the vehicle are detected by an automotive laser scanner and are used as landmarks.

1 Introduction

Many approaches for intelligent vehicles and active safety systems use environmental sensors in order to provide information about stationary and moving objects in the environment of a vehicle. Many algorithms depend on a correct determination of the change of the position and orientation of the host vehicle over ground between two points in time. This is indispensable for tracking algorithms, which are able to estimate the position and orientation as well as the speed over ground of tracked objects [1]. Furthermore, classification and situation assessment algorithms depend on this information [2].

The change of the position and orientation is estimated using dead reckoning algorithms. Therefore, the data of integrated yaw rate, steering angle and wheel speed sensors is combined. These algorithms perform well in standard situations, such as a drive on a highway or in urban areas.

However, in extreme situations, such as skidding or wheel spin on icy roads, dead reckoning algorithms based on these sensors are not able to measure the precise translation and rotation of the host vehicle, as the sensor data of the wheel speed encoders do not provide the precise movement of the vehicle. Tracking algorithms using this invalid information will not be able to estimate the correct direction and speed over ground of tracked vehicles. However, especially in these extreme situations it is indispensable for warning and active safety systems to obtain a correct representation of the environment, such as the correct direction, velocity and position of tracked objects.

Standard GPS is also not able to provide a precise information of the vehicle's movement in extreme situations, as the update rate of standard GPS-receivers is too low (1 Hz) and the orientation of the vehicle cannot be measured directly. Furthermore, there is a significant time delay between the point of time the position measurement was performed and the point of time the measurement is made available to the application [3].

Fig. 1. The laser scanner IBEO ALASCA XT is integrated at the front bumper of the testing vehicles.

In order to improve the determination of the change of the pose especially for tracking systems, stationary objects in the environment are used as landmarks. The landmarks are detected by an automotive laser scanner, which is mounted on the front bumper of the vehicle. Therefore, capable landmarks are detected using an advanced occupancy grid (online map). An algorithm for the robust association of detected landmarks to the distance measurements of the laser scanner is required for the determination of the movement of the vehicle in skidding situations.

2 Sensors

The multi–layer laser scanner ALASCA XT of the company IBEO Automobile Sensor GmbH acquires distance profiles of the environment of the vehicle of up to 270° horizontal field of view at a variable scan frequency. The angular resolution is chosen to 0.25° and the scan frequency is chosen to 10 Hz in this work. The laser scanner uses four scan planes with a vertical opening angle of 3.2°. The laser scanner is integrated in the front bumper of the testing vehicles as shown in Fig. 1. Furthermore, the data of the integrated yaw rate and steering angle sensors and the wheel speed encoders of the serial ESP system are used.

3 Algorithm Overview

This section gives an overview of the algorithms for the determination of the vehicle's movement between two laser scans. Fig. 2 shows the modules of the approach. The data of the integrated serial sensors of the vehicle, such as yaw rate, steering angle and wheel speed sensors, is combined in the movement estimation module. In this module, the change of the pose Δx_M, Δy_M, $\Delta\Psi_M$ of the host vehicle between two successive laser scans is estimated. In situations without wheel skin or skidding, this algorithm provides an accurate estimation of the vehicle's pose. These driving states are denoted to high adhesion maneuvers in this work.

The module online map generates an occupancy grid of the environment by accumulating distance profiles of the laser scanner in order to separate moving and stationary objects. The module landmark extraction determines capable landmarks from the online map in each time step. If the vehicle starts skidding, the movement estimation based on the integrated sensors will lead to inaccurate data. The situation is classified as a skidding situation in the module skidding detection using the yaw rate, cross acceleration and wheel speed sensors. The distance measurements of the laser scans are clustered to segments in the module segmentation. Landmarks, which are detected from the online map, are used in the localization module in order to determine the movement of the host vehicle during skidding maneuvers. The module landmark / dead reckoning switch provides either the results of the movement estimation module or the estimation of the landmark navigation to the application.

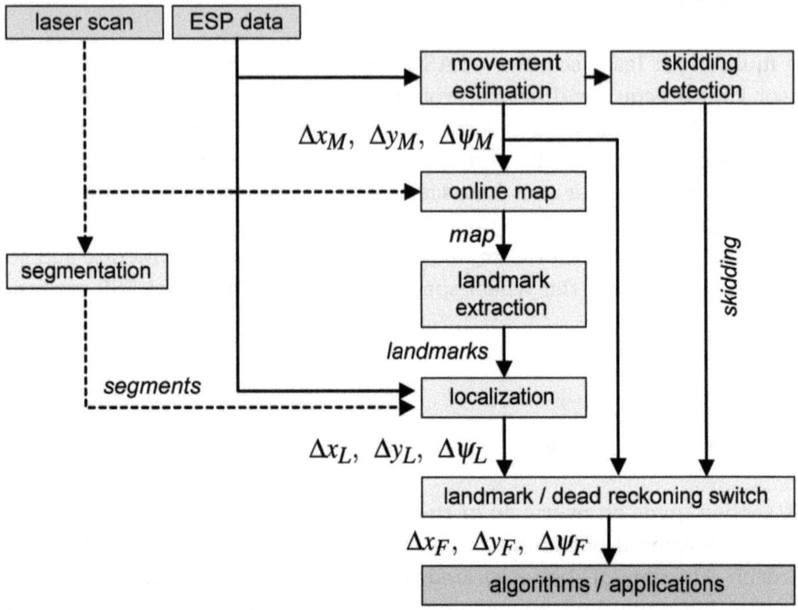

Fig. 2. Algorithm overview

4 Online Map

In this section, algorithms for the generation of online occupancy grids based on the distance measurements of the laser scanner are proposed. The algorithm is related to occupancy grid mapping, which is described in detail by Thrun et. al. [4, 5]. The basic idea is the partitioning of a certain region around the vehicle to grid cells and the accumulation of distance measurements in these grid cells. Each grid cell contains the history of the previous laser scans in form of a likelihood parameter, which specifies whether the cell is occupied or not.

In the following sections m_i denotes the cell i. All grid cells $m=\{m_i\}$ form the entire grid, which is denoted to online map in this work. $p(m_i)$ is denoted to the occupancy likelihood. The likelihood $p(m_i)=0$ stands for a free cell i and $p(m_i)=1$ stands for an occupied cell i [4].

4.1 Advanced Forward Inverse Sensor Model

In order to combine the current measurement and the existing online map, a separate grid with the same dimension as the online map is defined. The grid is denoted to measurement grid and it contains the occupancy likelihoods $p(m_j|z_t)$ in consideration of the current distance measurements of the laser scanner z_t at time t. m_j are denoted to the cells in the measurement grid.

4.1.1 Initialization of the Measurement Grid

All cells $k=1...N$ are set to $p(m_k|z_t)=0.5$, as it is not known whether the cell is occupied or not.

4.1.2 Transformation of the Distance Profile

The distance profile of the laser scanner is transformed to the map coordinate system in order to combine the cells of the measurement grid and the online map correctly.

4.1.3 Registration of Distance Measurements in the Measurement Grid

Each transformed measurement point is assigned to a certain grid cell in the measurement grid. There may be more than one distance measurements in some cells of the measurement grid, as the dimensions of the cells are chosen to 20 cm in this work. For each measurement point in a cell, the value 0.05 is added to the likelihood $p(m_j|z_i)$. The parameter was determined by extensive tests.

4.1.4 Registration of Free Regions in the Measurement Grid

Regions behind detected objects are neglected and the occluded cells keep their value 0.5, as shown in Fig. 3. There is information about free space in the distance profile of the laser scanner. The discrete angle steps of the laser scanner are known [7]. If the laser scanner does not detect an object at certain angle steps, the cells which are passed by the ray seem to be free. Furthermore, if the laser scanner detects an object, the cells which are passed by a line between the laser scanner and an object seem to be free. However, in far distances there may be small objects such as small posts, which are not detected by a laser beam of the actual scan. That's why the grid cells, which are situated along a laser beam, are not all set to zero, but to a value, which depends on the radial distance of the laser beam. For grid cells m_j, which are situated at a radial distance of d, the occupancy likelihoods of cells are calculated using the following equation:

$$p(m_j, d) = \begin{cases} \frac{0.4}{d_{\max}} \cdot d & \text{for } d \leq d_{\max} \\ 0.5 & \text{for } d > d_{\max} \end{cases} \quad (1)$$

The design parameter d_{max} is chosen with respect to the requirements of the application which uses the online map. In our application d_{max} is set to 50 m. That means that landmarks in a radial distance of 50 m are searched for. Fig. 3 shows the principle and two exemplary measurement grids of a highway and an urban scenario.

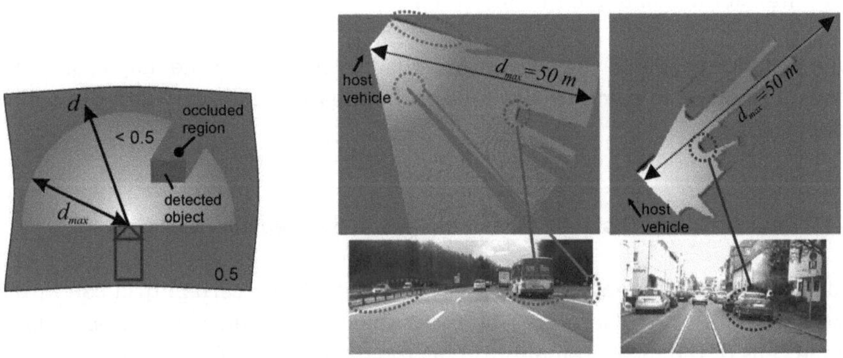

Fig. 3. Left: The likelihood of grid cells in the measurement grid depend on the radial distance to the laser scanner.
Right: Exemplary measurement grids of a highway and an urban scenario.

4.2 Update of the Online Map

In our case, the occupancy of a grid cell is estimated. The cells remain at the same position and it is assumed that the occupancy of cells does not change during sensing. Binary Bayes filters address these estimation problems [4]. The current online map cells $p(m_i|z_1,\ldots,z_t)$ are determined from the existing online map of the last time step $p(mi|z_1,\ldots,z_{t-1})$ and the actual measurement grid $p(m_i|z_t)$ using Bayes Theorem:

$$p\left(m_i \mid z_1,\ldots,z_t\right)=\frac{p\left(z_t \mid z_1,\ldots,z_{t-1},m_i\right)\cdot p\left(m_i \mid z_1,\ldots,z_{t-1}\right)}{p\left(z_t \mid z_1,\ldots,z_{t-1}\right)} \tag{2}$$

The measurement in a cell m_i only depends on the measurement z_t, because the cells of the measurement grid are all set to 0.5 before distance measurements of the actual lasers scan are registered:

$$p\left(z_t \mid z_1,\ldots,z_{t-1},m_i\right)=p\left(z_t \mid m_i\right) \tag{3}$$

The likelihood for a free cell $p(m_i=0|z_1,\ldots z_t)$ and an occupied cell $p(m_i=1|z_1,\ldots z_t)$ are divided. With respect to $p(\bar{A})=1\text{-}p(A)$ and $p(\bar{A}|B)=1\text{-}p(A|B)$ the odds ratio is calculated, which gives the likelihood of the cells in the online map in time step k:

$$p\left(m_i \mid z_1,\ldots,z_t\right)=\frac{S}{S+1} \tag{4}$$

$$S=\frac{p\left(m_i \mid z_t\right)}{1-p\left(m_i \mid z_t\right)}\cdot\frac{p\left(m_i \mid z_1,\ldots,z_{t-1}\right)}{1-p\left(m_i \mid z_1,\ldots,z_{t-1}\right)} \tag{5}$$

Truncation problems near 0 and 1 are considered by using a simple procedure. If a likelihood of a grid cell is $p(m_i|z_1,\ldots z_t)>\varepsilon$ or $p(m_i|z_1,\ldots z_t)<\varepsilon$, then the cells are set to $p(m_i|z_1,\ldots z_t)=1\text{-}\varepsilon$ or $p(m_i|z_1,\ldots z_t)>\varepsilon$, respectively. Fig. 4 shows an example.

Fig. 4. Exemplary measurement grid and online map of an urban scenario. The red points in the online map are the actual distance measurements of the laser scanner.

4.3 Vehicle Movement in the Online Map

The change of the position and the orientation of the vehicle (movement) are determined from serial ESP sensors. The yaw rate, steering angle and wheel speed encoders are combined using a Kalman filter. A bicycle model is used in the prediction step.

The online map is only updated in high adhesion maneuvers. In the first step the change of the pose Δx_M, Δy_M, $\Delta\Psi_M$ of the vehicle are determined from the integrated ESP-sensors, which provide satisfying data concerning the accuracy in these maneuvers. The number of columns and rows, which were passed by the vehicle, is calculated. As only a quadratic region (the online map) is held in the memory of the computer, cells are added and removed as shown in Fig. 5.

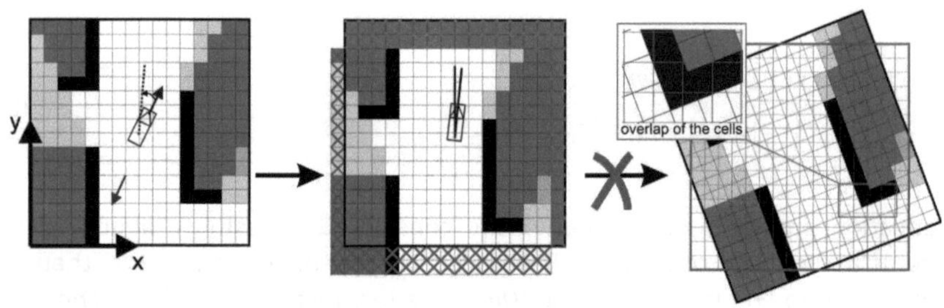

Fig. 5. Movement of the host vehicle over the online map: Cells are added and removed. The vehicle is turning relative to the online map.

The vehicle is turning relative to the map. Thus, discretisation errors which would occur if the map is turned relative to the vehicle are eliminated (Fig. 6).

4.4 Extraction of Landmarks from the Online Map

The main benefit of the online map is the robust separation of moving from stationary objects. That's why the online map is used to determine valid stationary landmarks in the environment of the host vehicle. The determination of landmarks is performed in three steps:

The online map is converted to a binary image. A threshold for the likelihoods of the cells in the online map is used for the decision, whether a cell is added to the binary image or not. A contour detection algorithm clusters the binary image. The clusters are analyzed. Those clusters, which show a valid dimension and a valid shape, are stored as landmarks.

In this approach, valid landmarks are small objects such as posts of traffic lights and signs. Furthermore, the approach can be extended for other types of objects, such as corner or line shapes. Fig. 6 shows two exemplary scenes. The left scenario is a free testing field, where several posts were placed. This test scenario was used to analyze the landmark navigation and tracking approaches. The right example is an urban scenario. Extensive tests have shown, that point and corner landmarks can be found in most urban and highway scenarios. The real time landmark extraction algorithm is performed in every time step.

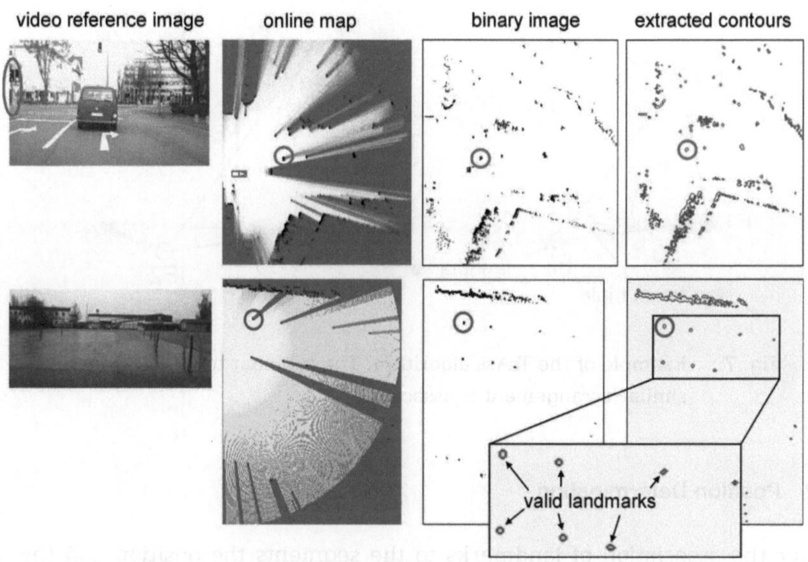

Fig. 6. Extraction of landmarks from the online map: Video reference images show the scenario. A binary image is generated from the online map and valid landmarks are extracted from contours in the binary image.

5 Localization

5.1 Landmark Association

In order to determine the vehicle's position relative to the extracted landmarks, an algorithm for the association of extracted landmarks to the segments in the clustered laser scan is required. Therefore, the algorithm TrAss (TrAss - Triangle Association) is used, which is introduced in detail in [7]. The basic idea of the algorithm is the definition of a translation and rotation invariant representation of geometric arrangements using triangles. Triangles are completely characterized by the lengths of their three side lengths and the side lengths of a triangle do not depend on the position and the orientation of the triangle. A set of three landmarks represents the vertices of a triangle. Fig. 7 shows an example. In each time step landmarks are associated. TrAss is able to deal with the rapidly changing environment.

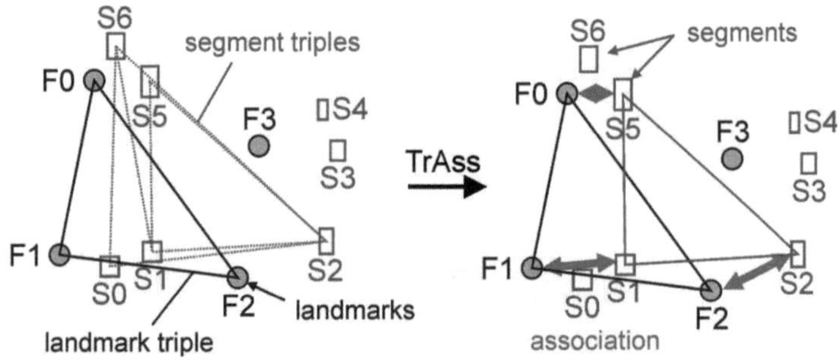

Fig. 7. Example of the TrAss algorithm. The segment triple with the most similar arrangement is associated.

5.2 Position Determination

After the association of landmarks to the segments the position and the orientation of the host vehicle is determined by comparing the segment and landmark arrangement using a point matching algorithm. For n-pairs of points, which represent the associated segments and landmarks, a function E for the distance between the points $\mathbf{p}_i$ representing the coordinates of the segments and the points $\mathbf{p}'_i$ representing the coordinates of the landmarks is minimized [7,8].

$$E = \sum_{i=1}^{n} \left(\mathbf{R}(\varphi) \cdot \mathbf{p}_i + \mathbf{t}_i - \mathbf{p}'_i \right) \tag{6}$$

The minimization step determines the translation $\mathbf{t}_i=[\mathbf{t}_X, \mathbf{t}_y^T]$ and the orientation angle ψ with respect to the rotation matrix $\mathbf{R}(\psi)$. The translation and rotation gives the pose of the vehicle relative to the landmarks during skidding.

6 Environment Perception

Several advanced driver assistant systems need information about objects in the vehicle's environment. A consistent environment description can serve multiple assistant systems simultaneously. The environment perception algorithms provide an object list, containing information like position, size, velocity and class of all measured objects. First, a segmentation algorithm groups close measurements. Details about this segmentation can be found in [9]. Afterwards, a Kalman filter based tracking is applied. Details about the tracking algorithm and its dependence on a good estimation of the host vehicle's motion will be given in the next section. Finally, an object classification is performed based on the measured features, the dynamic state and additional knowledge like precise maps [10]. The system layout is illustrated by Fig. 8.

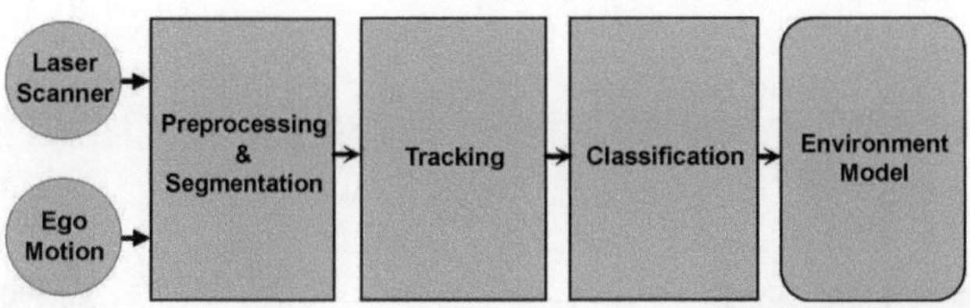

Fig. 8. Layout of the environment perception algorithms

7 Tracking

The applied tracking is based on algorithms, which were proposed by [1]. The tracking applies a Kalman filter to each of the measured objects. The dynamic model is a free mass model. The filters estimate absolute object velocities.

7.1 Coordinate System and Ego Motion Compensation

All measured objects are tracked in the host vehicle's coordinate system. The objects position is usually measured with a sensor frequency of 10 Hz. The sensor measures different positions at different time steps, which correspond to the object's velocity. Unfortunately, this process is caused by a superposition of the host vehicle's motion and the real absolute velocity of the measured object. In order to estimate absolute object velocities, a compensation of the host vehicle's motion is necessary. This compensation is applied at each time step of the Kalman filter after the prediction of the expected object position. The predicted position of each object is moved by the inverse of the estimated ego motion. The uncertainty of the ego motion is incorporated into the covariance matrix of each processed object.

This approach can handle small errors of the estimated ego motion. Nevertheless, extreme errors will lead to extreme errors of the predicted object positions. These errors will inhibit a correct association of new measurements to the processed object tracks. Hence, the object tracks will fail or even be lost.

Usually, the error of the estimated ego motion which is calculated based on the vehicles onboard motion sensors, is quite small. Exceptional circumstances like sliding of the host vehicle can jeopardize the complete tracking algorithms, if the host vehicles motion estimation is too erroneous. For this reason, the new approach of the ego motion estimation is essential to guarantee the operability of the tracking algorithms.

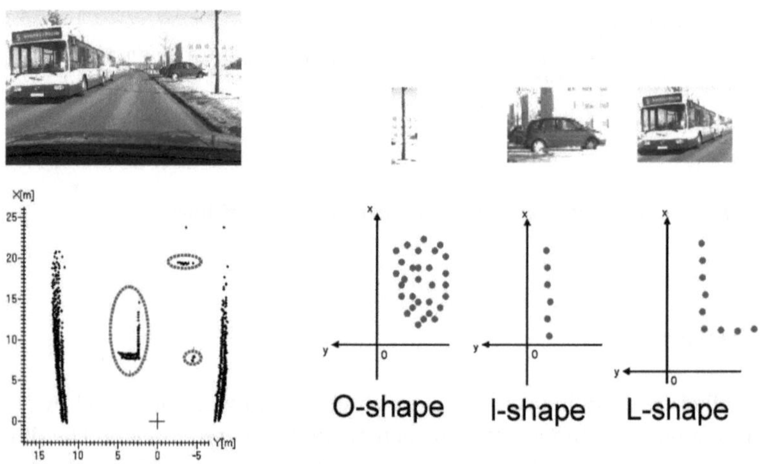

Fig. 9. Three types of shape are distinguished. An L-Shape is typical of rectangular objects, measured from two sides. The I-Shape is the characteristic shape of walls, fences or rectangular objects, which are only measured from one side. An O-shape is assigned to the remaining objects.

7.2 Feature Extraction

The laser scanner measures the objects' positions as well as contour information due to the sensors high angular resolution. The feature extraction calculates a shape, which is fitted into the laser scanner measurements. Three types of shape are distinguished (Fig. 9). If the laser scanner measures two sides of a rectangular object, a L-shape will be fitted into the measurements. If only one side is measured, an I-shape will be used. An O-shape is assigned to the remaining objects [9].

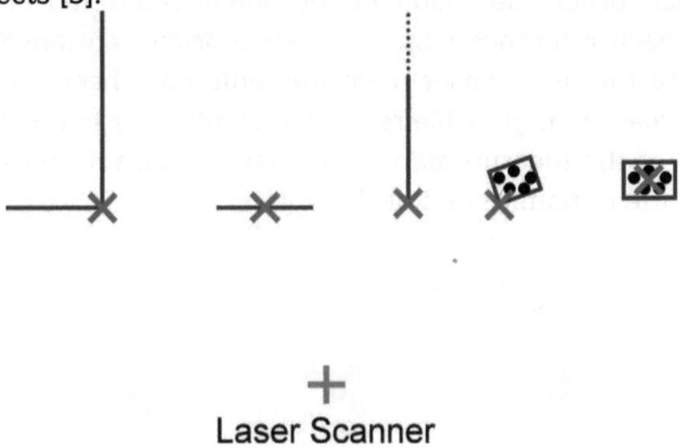

Fig. 10. Different reference points are chosen, depending on the assigned object shape. The reference point corresponds to an edge of the object, to the middle of an object side or to the centre of geometry.

Different reference points are chosen, depending on the estimated object shape (Fig. 10). The corner is selected for each L-shape. If the complete side of an I-shape is observable, the middle of the side is chosen. If one corner of an I-shape is not measured, due to occlusions or the inappropriate orientation to the sensor, the measured corner will be selected as reference point. If an O-shape is assigned, it is not possible to measure the orientation. There are two possibilities to handle an O-shape. If the orientation was measured before, the predicted orientation is assumed. A bounding box with this orientation is calculated, which contains all measurements. The appropriate reference point is chosen depending on the orientation to the sensor. If the orientation was not measured before, an orientation of zero degree is assumed. A corresponding bounding box is calculated and the center of geometry is used as reference point.

7.3 Object Model

A rectangular object model is applied to approximate the measured objects. Consequently, the dynamic state of each object is given by position, size, orientation, yaw rate, velocity and acceleration. Usually, the measured shape not only depends on the object, but also on its orientation to the sensor. For this reason, different reference points can be used to describe the object's position at different time steps. Fig. 11 illustrates the positions of possible reference points. An object corner, the middle of one side or the center of geometry corresponds with each reference point. The appropriate reference point can not be selected until the new sensor measurements have been associated to the object. For this reason, all possible reference points have to be predicted. After the association of the measurements, the appropriate reference point will be used to update the dynamic state of the object.

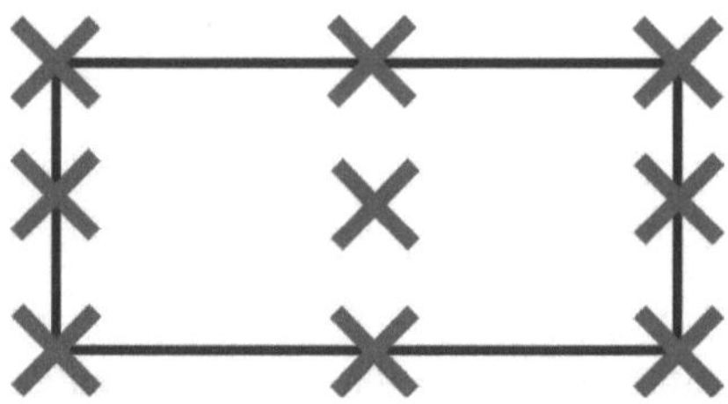

Fig. 11. Possible reference points: Depending on the current sensor measurements, different reference points will be used to describe the position of the object. The reference point can correspond with an object corner, the middle of an edge or the geometric center of the object.

7.4 Kalman filter Overview

The dynamic state estimation is performed with a Kalman filter [11] for each of the measured objects. Fig. 5 gives an overview about the algorithm. The layout is quite similar to the layout proposed in [1], but some small changes were applied to improve the processing time.

The new measurements are obtained from the laser scanner. The segmentation groups the measurements and calculates features like shape and position.

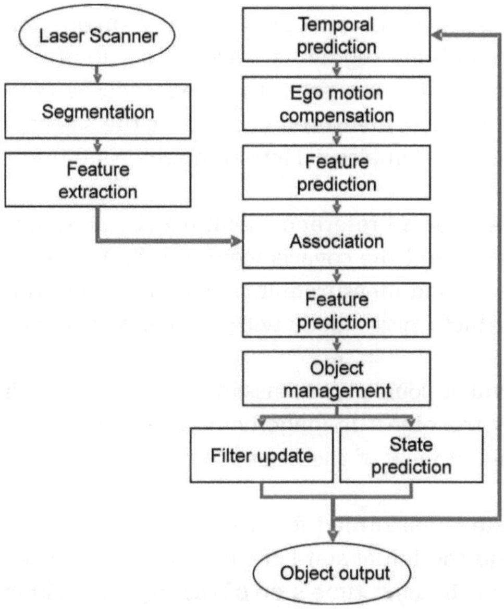

Fig. 12. Layout of the Kalman filter based tracking algorithm

The steps of the Kalman filter are given in Fig. 12 on the right. First, the dynamic state of the objects is predicted to the current time. Afterwards, the motion of the host vehicle is compensated. Now all features have to be predicted. Since the appropriate reference point is not known, all possible reference points have to be predicted. If this is performed directly by the measurement matrix of the Kalman filter in (7), this matrix will have a quite large dimension.

$$\mathbf{z}_k^- = \mathbf{H} \cdot \mathbf{x}_k^- \tag{7}$$

Consequently, the matrix inversion of the Kalman filter update step (8) will have to be performed with a matrix of quite large dimension.

$$\mathbf{x}_K = \mathbf{x}_k^- + \mathbf{P}_k^- \mathbf{H}^T \left(\mathbf{H}\mathbf{P}_k^- \mathbf{H}^T + \mathbf{R}\right)^{-1} \left(\mathbf{z}_k - \mathbf{H}\mathbf{x}_k^-\right) \tag{8}$$

With respect to the necessary processing time, the inversion of a large matrix is not advisable. Therefore, the applied approach uses the measurement matrix only to predict one reference point.

After the ego motion compensation the center of geometry is predicted with the known measurement prediction in (8). Afterwards, the positions of the remaining reference points can be calculated by incorporating the estimated object's dimensions and orientation.

The data association selects appropriate measurements for the objects. It is possible to assign multiple segments with respect to given conditions [10]. If more than one segment is assigned, the shape will be recalculated based on all assigned laser scanner measurements. After the selection of the measurements, the data association also selects an appropriate reference point.

Unfortunately, the selected reference point usually does not correspond to the measurement matrix and the covariance of the first measurement prediction. For that reason, a second measurement prediction is performed to predict the reference point, which corresponds with the measured one.

Measurements, which could not be assigned to tracked objects, are assigned to new objects by the object management. This part also deletes old objects, which left the field of view of the laser scanner.

Now a filter update is performed for tracked objects and a state prediction is applied to calculate the initial state of the new objects. The new state estimation is then used at the next time step of the object tracking.

Experiments have shown that the proposed reduction of the measurement matrix and the corresponding matrices reduces the necessary processing time of the Kalman filter steps to a third.

8 Results

The algorithms were tested on a free testing field with a low adhesion surface. A virtual street was added for visualization. One vehicle is driving through the scenario with an almost constant velocity of 11 m/s. Another vehicle, which is equipped with a laser scanner, is following and tracking the proceeding vehicle. Several skidding maneuvers were performed by the second vehicle and the position accuracy of the skidding vehicle using landmark navigation was determined using a RTK-GPS receiver. The position error of the positioning algorithm using extracted landmarks is in the range of 20 cm to 40 cm depending on the scenario. Fig. 13 shows the results of a drift maneuver. The vehicle is moving across the longitudinal axis. In these driving situations, the position and orientation of the vehicle can not be determined using the serial ESP sensors. Tracking systems are able to determine the ground truth movement of other vehicles even in these hard conditions. Extensive tests have shown that even in urban and highway scenarios, many valid landmarks for the approach can be expected and the approach will also work in these scenarios. Furthermore, the system is applied in full braking scenarios.

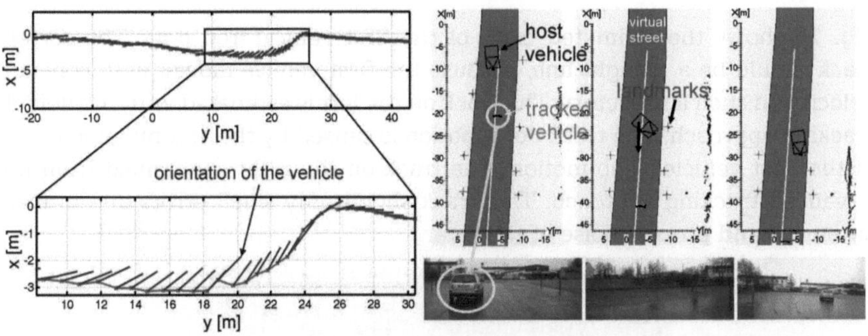

Fig. 13. The host vehicle is skidding (drift manoeuvre, left turn then a fast right turn). A virtual street was added for visualization. The orientation and position of the vehicle to a reference coordinate system is determined precisely using extracted landmarks. The laser scan (black points) and the extracted landmarks (red crosses) are shown.

Fig. 14 illustrates the improvement of the advanced tracking system using the localization approach. The estimated velocity of the first vehicle is plotted. If the motion of the second vehicle (host) is measured by onboard ESP sensors, the estimated velocity of the first vehicle is very erroneous during the skidding maneuver (blue line). The calculation of the second vehicle's motion with the new landmark approach can significantly improve this estimate (red line).

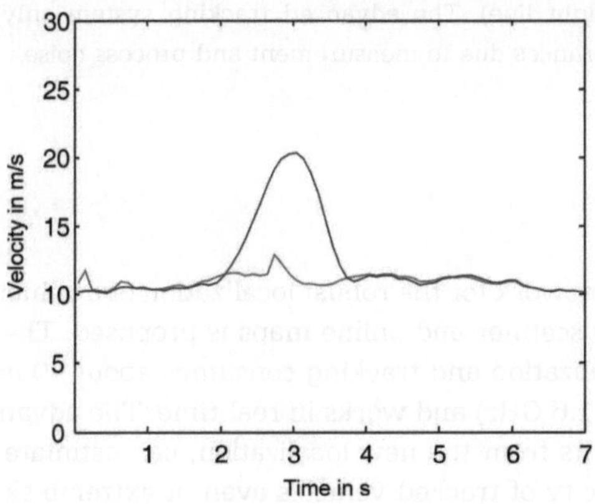

Fig. 14. Improvements of the advanced tracking system: The blue line shows the estimated velocity of the first vehicle for the default tracking. The red line shows the estimated velocity of the advanced tracking system with improved, landmark based calculation of the host's motion.

Fig. 15 shows the estimated track of the first vehicle (red box). The correct track should be a straight line, because the first vehicle moved with constant velocity on such a trajectory. The track on the left is estimated with the default tracking approach. The track deformation is caused by the accumulated errors of the host vehicle's ego motion. The track on the right is obtained from the advanced tracking approach. This track shows only small errors due to measurement and process noise of the filter.

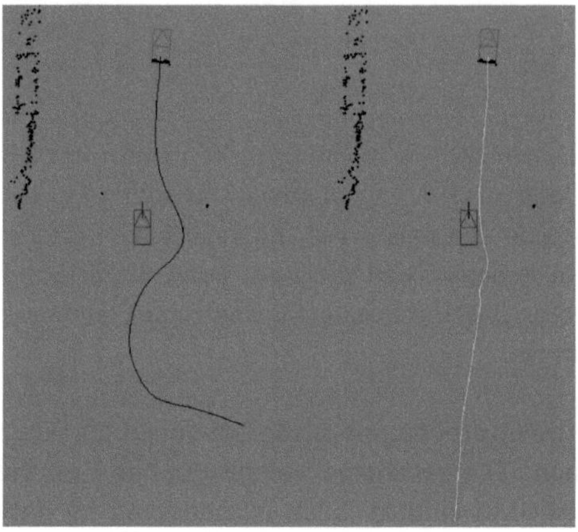

Fig. 15. Estimated track of the first vehicle: The estimated track of the default system significantly differs from the correct trajectory (straight line). The advanced tracking system only shows small differences due to measurement and process noise.

9 Conclusion

An algorithm framework for the robust localization of a vehicle in skidding scenarios using laser scanner and online maps is proposed. The entire algorithm including the localization and tracking consumes about 40 ms of computation time (Pentium IV 1.6 GHz) and works in real time. The advanced tracking system, which benefits from the new localization, can estimate the position, orientation and velocity of tracked vehicles even in extreme skidding situations. These results will improve future safety systems and situation assessment in extreme situations.

10 Acknowledgements

Please visit www.argos.uni-ulm.de for more information about our research projects. This work is supported by IBEO Automobile Sensor GmbH (www.ibeo-as.de).

References

[1] N. Kaempchen, M. Buehler, K. C. J. Dietmayer: "Feature-Level Fusion for Free-Form Object Tracking using Laserscanner and Video", Proceedings of 2005 IEEE Intelligent Vehicles Symposium, Las Vegas, USA, June 2005.

[2] S. Wender, T. Weiss, and K. Dietmayer: "Improved object classification of laserscanner measurements at intersections using precise high level maps," Proceedings of the 8th International IEEE Conference on Intelligent Transportation Systems, Vienna, Austria, 2005.

[3] T. Weiss, S. Wender, and K. Dietmayer: "Praezise Positions- und Zustandsschaetzung eines Fahrzeugs mit Hilfe eines Laserscanners und detaillierten digitalen Karten. (Precise Position and State Estimation of a vehicle using laserscanners and detailed digital maps)", 4th Workshop on Driver Assistant Systems, Loewenstein, 2006.

[4] S. Thrun, D. Fox, and W. Burgard: "Probabilistic Robotics". MIT Press, 2005.

[5] S. Thrun: "Learning occupancy grid maps with forward sensor models", [Online]: Available: citeseer.ist.psu.edu/701378.html.

[6] IBEO-AS GmbH Website, www.ibeo-as.de, 2007.

[7] T. Weiss, N. Kaempchen, and K. Dietmayer: "Precise ego localization in urban areas using laserscanner and high accuracy feature maps", Proceedings of 2005 IEEE Intelligent Vehicles Symposium, Las Vegas, USA, 2005.

[8] Feng Lu, Evangelos Milios: Robust Pose Estimation in Unknown Environments by Matching 2D Range Scans, Department of Computer Science, York University, North York, Canada, 1994.

[9] Wender, K. Fuerstenberg, K. C. J. Dietmayer: "Object Tracking and Classification for Intersection Scenarios Using A Multilayer Laserscanner", Proceedings of ITS 2004, 11th World Congress on Intelligent Transportation Systems, Nagoya, Japan, October 2004.

[10] S. Wender, T. Weiss, K. Fuerstenberg, K. C. J. Dietmayer: "Object Classification exploiting High Level Maps of Intersections", Proceedings of Advanced Microsystems for Automotive Applications 2006, Berlin, Germany, April 2006.

[11] Y. Bar-Shalom, X. Li: "Estimation and Tracking: Principles, Techniques, and Software", Artech House, London, 1993.

Thorsten Weiss, Stefan Wender, Klaus Dietmayer
Institute of Measurement, Control and Microtechnology, University of Ulm
Albert-Einstein-Allee 41
89081 Ulm
Germany
thorsten.weiss@uni-ulm.de
stefan.wender@uni-ulm.de

Keywords: landmark navigation, ego motion, skidding maneuver, tracking, laser scanner

Fast Fusion of Range and Video Sensor Data

A. Linarth, J. Penne, B. Liu, O. Jesorsky, Elektrobit Automotive Software
R. Kompe, Friedrich-Alexander University, Erlangen-Nuremberg

Abstract

This paper brings an innovative approach to generate 3D scene views from real world data using a matrix range sensor and a 2D video sensor. Besides the enhanced visualization, combining 2D and 3D data provides also valuable information for object detection and recognition as well as for other image processing methods. The scope of this work includes a discussion on extracting good features from low resolution images, necessary for the system calibration. The registration process is based on a geometrical model which is derived from the calibration parameters of the cameras. Finally, a FPGA-based solution is provided, which generates the 3D scene on-the-fly. The proposed architecture achieves its maximum performance with a fully pipelined implementation from the raw data to the generated fusion coordinates, while limited hardware resources are consumed. In contrast to standard processor approaches, a significant performance boost is achieved with the FPGA architecture.

1 Introduction

Photorealistic 3D models are a desired feature in many computer vision related applications, e.g. virtual and augmented reality. The proposed system takes advantage of the new generation of 3D sensors based on the time of flight principle which provides individual distance information for each element of a sensor matrix besides intensity and amplitude values [6, 3]. Although the available resolution of these sensors is sufficient for a bunch of image processing methods, current lateral resolutions cannot yet compete with standard 2D video sensors. Relating the low-resolution depth frame to a standard 2D monochrome or color image enhances the results of many image processing methods, such as segmentation, tracking or classification. The complexity level of these operations can also be reduced, which simplifies their use in embedded platforms. The described 2D-3D fusion follows the increasing interest of the automotive industry on imaging systems, with applications that range from pedestrian recognition to parking aid systems, as well as in the manufacturing process like quality inspection.

In the following of this paper a discussion about the pattern and the feature extraction used in the calibration procedure is firstly presented (Section 2). Then a geometrical model is derived for the rigid registration between two camera coordinate systems (Section 3). An efficient scheme for the fusion system using FPGA is shown afterwards (Section 4), followed by an experiment analysis of this work (Section 5).

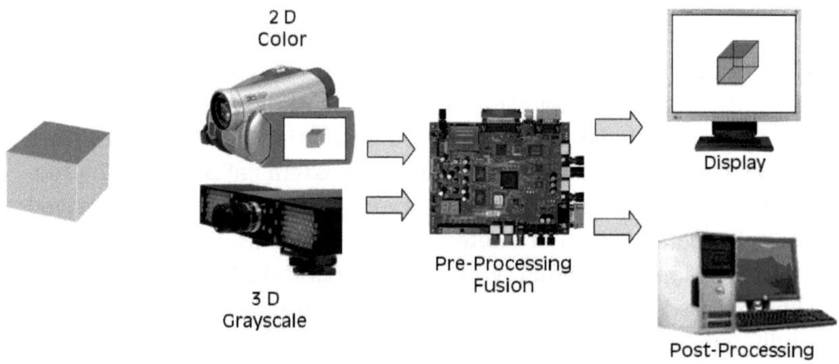

Fig. 1. System overview

2 Camera Calibration

The geometrical approach implemented for the fusion process requires a reliable estimation of the cameras positions and their internal parameters. The calibration is performed by using the method presented by Zhang [7] for both cameras simultaneously, which makes it possible to estimate the relative position and orientation between the two cameras at the same time. To choose a good calibration pattern and to determine the most accurate point correspondences are essential conditions for achieving good results. The intensity images with a low resolution of 64x16 supplied by the range sensor and the color images with a high resolution of 720x576 supplied by the 2D video sensor are treated in different manners in the pre-processing phase.

2.1 Calibration Pattern

Choosing a proper calibration pattern was an important issue during the development process due to the low-resolution feature of the time-of-light camera. A minimum of 4 point correspondences is a prerequisite imposed by the homography calculation used in the Zhang method. A circle based pattern was chosen instead of standard chess boards, since the latter shows higher sensibility to the infra-red lighting condition supplied by the 3D camera. Corners and intersections were by far visibly harder to be identified than the circles' centroids (gravity centers), and increasing the number of squares consequently reduces their sizes making even harder to identify good correspondence points. Fig. 2 shows images from these two patterns.

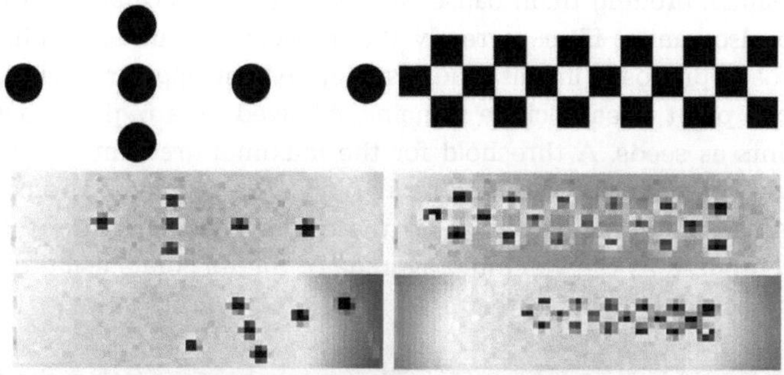

Fig. 2. Video images and range images of the circle pattern and chessboard pattern

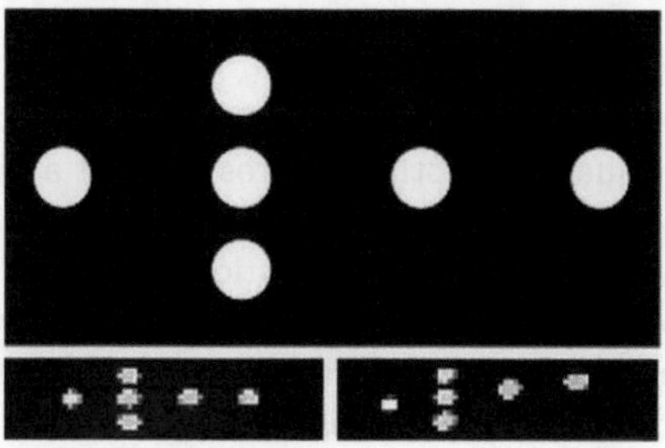

Fig. 3. Final calibration pattern

Concerning the geometrical relation between both cameras, to improve the precision of the derived model, each point is registered to the second image applying the registration procedure. This method takes the advantage of the depth data grabbed by the 3D range sensor at the given position. Therefore, a final calibration pattern with white circles on black ground was chosen since the low reflectance of the black material yields unreliable distance measurements. The final calibration pattern is shown with two of its 64x16 intensity images from the 3D camera.

2.2 Point Correspondences

The calibration process is based on the point correspondences between the extracted circular feature points on the images and the corresponding ones on the actual calibration pattern. Applying a simple threshold followed by erosion is enough for separating the interesting regions in the high resolution video image. However, this process does not provide satisfying result for the low resolution grayscale depth image from the 3D camera, since the circles are very small, eroding them causes a great loss of information. A minimum threshold also cannot filter correctly the information due to lighting effects. The approach proposed in this study is to apply a strong threshold, such that at least one point of each circle remains, followed by a region growing with these points as seeds. A threshold for the maximal gradient change is used as the stop condition for the region growing. The segmented region is then considered as the effective region for the circle. The centroid of the circle is further computed as the weighted average of the its points using their intensity values as the weights.

Identifying a circle in a calibration pattern can be done using the following property:

$$d = \frac{(circumference)^2}{area} = \frac{(2\pi r)^2}{\pi r^2} = \frac{4\pi^2 r^2}{\pi r^2} \approx 12.6 \quad (1)$$

The ratio between the squared circumference and area of a circle should be equal to the above value, while other figures present a bigger value. In practice, an error level is set, since circles get projected as ellipses. Although this property works so well with the high resolution image, for a 64x16 pixel image a circle can be represented with 10 pixels or even fewer, sometimes with formats that can be easily confused with squares or rectangles. The centroid of such a small circle is also not precise enough for the calibration method. Therefore, adjusting the error level can represent a good method for determining if the circle is good/big enough for the calibration process. In this paper no further studies were done in this direction, but an empirical error level was chosen based on experiments. Fig. 4 shows a reversed image of the calibration pattern with the circle centers marked after ordering them.

2.3 Calibration Procedure

- Place the calibration pattern such that it can be clearly sensed by both cameras and then synchronously acquire an image pair.
- Move the pattern such that the pattern plane is not coplanar to those where pictures have already been taken (see Zhang [7]) and take another pair of images. Repeat this step at least 3 times. The more pictures are taken, the higher is the accuracy of the result.
- Calculate point correspondences for each of the images.
- Calculate internal and external parameters using the Zhang method.
- Derive the relative transformation between both cameras and the registration matrix from one of the image pairs. (see the following section)

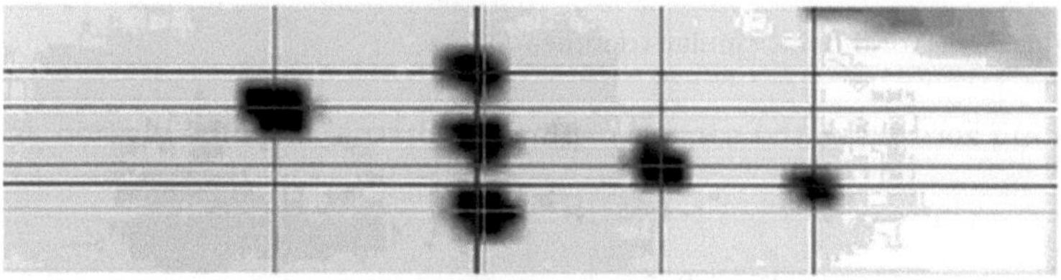

Fig. 4. Estimated centroids of the circles on a low-resolution image

3 Computation of the Fusion Coordinates

Assuming a rigid configuration of a 3D and a 2D camera, viewing to the same scene, a geometrical model can be found to register the 3D information of time of flight sensor with the 2D information of the video camera. Compared to stereo methods, the approach proposed here does not need feature points to identify correspondences. Instead, with previously calibrated camera parameters and the relative transformation between the two camera, it is only necessary to calculate the 3D world points from the sensed distance values using the camera parameters achieved from the calibration procedure and to project them into the 2D image.

Taking advantage of the distance d_w from the centre of the time of flight camera to the object, the back projection to world points, can be derived using triangulation, as seen in Fig. 5, where d_i represents the distance from the center of the camera to the image point, w the pixel width, h the pixel height and f the focal length.

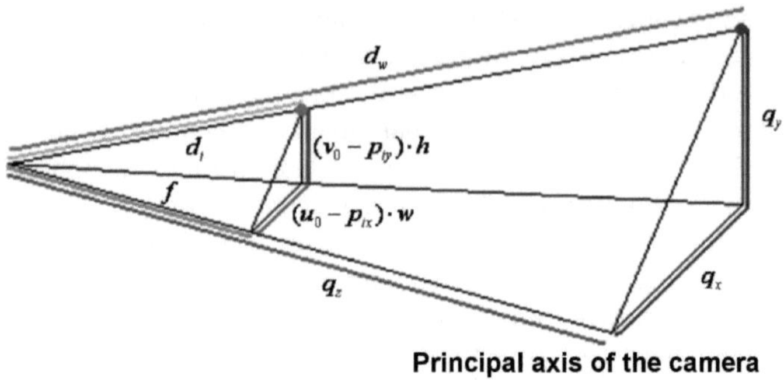

Fig. 5. World point similar triangles

Assuming zero skew, the camera calibration matrix K can be written as [2]:

$$K = \begin{bmatrix} \alpha & \gamma & u_0 \\ 0 & \beta & v_0 \\ 0 & 0 & 1 \end{bmatrix} = \begin{bmatrix} \frac{f}{w} & 0 & u_0 \\ 0 & \frac{f}{h} & v_0 \\ 0 & 0 & 1 \end{bmatrix} \tag{2}$$

Scaling the image triangle with $1/f$, and denoting as d'_i the scaled distance of the image point to the camera, the world point (q_x, q_y, q_z) can be calculated by:

$$s_x = \frac{u_0 - p_{ix}}{\alpha} \tag{3}$$

$$s_y = \frac{v_0 - p_{iy}}{\beta} \tag{4}$$

$$d'_i = \sqrt{\left(s_x^2 + s_y^2 + 1\right)} \tag{5}$$

$$q_x = s_x \cdot \frac{d_w}{d'_i} \tag{6}$$

$$q_y = s_y \cdot \frac{d_w}{d'_i} \tag{7}$$

$$q_z = \frac{d_w}{d'_i} \tag{8}$$

Once the world points are calculated, the next step is to project it back to the image plane of the video camera. At this point the coordinates of the world points are expressed under the coordinate system of the time-of-flight camera, it is necessary to transform the coordinate to the coordinate system of the video camera (see Fig. 6).

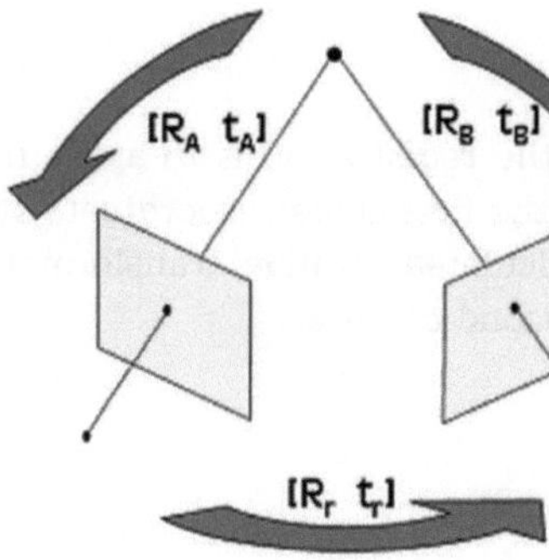

Fig. 6. Relative transformation

The transformation between the coordinate systems can be calculated by relating the external parameters of them at each scene and it is constant in a rigid configuration. In the following equations footnotes *A* and *B* are added to distinguish the two cameras, and letting m be the 3D world coordinates in homogeneous coordinates. Let the image points p_i [2]:

$$p_{iA} = K_A [R \quad t]_A m \tag{9}$$

$$p_{iB} = K_B [R \quad t]_B m \tag{10}$$

Let:

$$q = K_A^{-1} p_{iA} = [R \quad t]_A m \tag{11}$$

Separating rotation *R* and translation *t* components:

$$m = R_A^{-1} (q - t_A) \tag{12}$$

Hence:

$$p_{iB} = K_B [R \quad t]_B R_A^{-1} (q - t_A) \tag{13}$$

$$p_{iB} = K_B \left(R_B R_A^{-1} q - R_B R_A^{-1} t_A + t_B \right) \tag{14}$$

Interpreting q as a world point whose coordinate system's origin is at the centre of the first camera, the relative transformation between both coordinate systems is given by:

$$R_r = R_B R_A^{-1} \tag{15}$$

$$t_r = t_B - R_B R_A^{-1} t_A \tag{16}$$

Therefore the final step on the registration is to apply the projection of the world points with the origin the first camera coordinate system to the second camera, by applying the calculated relative transformation and finally the internal parameters of the second camera:

$$p_{iB} = K_B \begin{bmatrix} R & t \end{bmatrix}_r \begin{bmatrix} q \\ 1 \end{bmatrix} \tag{17}$$

where $q = (q_x, q_y, q_z)$.

4 On-the-Fly Implementation on FPGA Experiments

Developed for the Xilinx Virtex II 3000 FPGA [Xil06] available in the Celoxica RC203-E development kit [Cel06], the realized system targets the optimization of data-flow and the parallelization of processes to achieve a fast fusion method for the 3D data of the time of flight system and the 2D video data. The system developed in this work can be roughly described by the data flow between the three modules, shown in Fig. 7.

While the input and output processes are responsible to interface the external devices, the core of the implementation is represented by the fusion module. The input and output modules were developed to provide an interface to the camera and to a display, respectively. In the final system, the depth calculation procedure can be connected with a very simple interface to the analog to digital converter, and the output module would be connected to further image processing modules.

The fusion module was developed as a fully pipelined architecture. It receives the raw pixel data in a progressive scan fashion and provides at the same frequency depth, grey scale, amplitude of the infrared light as well as the relative coordinates of such points in the second image.

In the equations to calculate q_x, q_y and q_z it is possible to identify that just the depth gets changed at a given time. All other values are constant for each pixel while the configuration of the camera doesn't change. Therefore, a look-up table can store pre-calculated values, reducing the calculation of the world points to three simple multiplications which can be done in parallel, wasting one single clock cycle. Fig. 8 shows the hardware architecture for this module. The look-up tables are stored in dual-port RAMs. This is an efficient approach since one of the ports can be connected to a configuration module (ex. a processor), while the second one is read in a sequential fashion, to provide the coefficients at each point, avoiding time multiplexing, except for the fact that multiple access to the same address should be avoided (combinatively implemented in "read after write" mode). The table should contain one entry for each pixel on the image, which means 64x16 = 1024 entries. By choosing a 32bit representation, each of the three look-up tables can be implemented using 2 Block-RAMs, out of 96 in total of the given FPGA. To connect to the system, a configuration module, representing a core generated OPB bus memory interface was adapted to connect to the described dual-port RAMs.

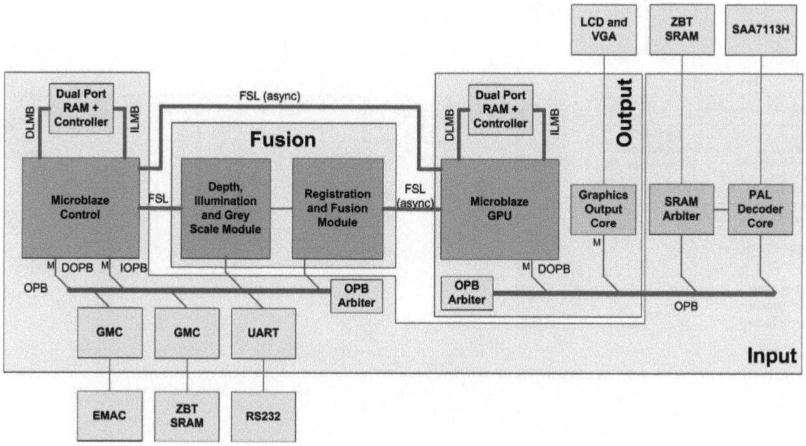

Fig. 7. FPGA implementation architecture

Several approaches could be used to solve the registration equation in hardware. One of them could be pre-multiply intrinsic an extrinsic parameters and reducing the operation to a simple matrix multiplication. This multiplication, by its time, could be solved, for example, sharing just one adder and one multiplier in a MAC fashion. This could be an interesting method if the data arrives in 32bit words, in a sequential way. However, in the present case, world points are available at a single clock cycle, allowing a more efficient approach to be implemented. The bottom part of Fig. 8 shows the proposed architecture for this matrix multiplication.

Applying this method, an on-the-fly computation is achieved. Multiplication of *X*, *Y* and *Z* coordinates are done in parallel and stored in intermediate registers (in the diagram, all the operations present an internal register at the output). At the two next levels, these registers are added, providing the result for the matrix multiplication in just 3 clock cycles. 2 or 1 clock cycles would also be possible but to reduce the overall latency, this is avoided. Since the coordinates were until now represented homogeneously, this is a good time to dehomogenize them. Therefore two core generated pipelined dividers were added, whose results give the coordinate on the second image of the respective point of the time of flight camera.

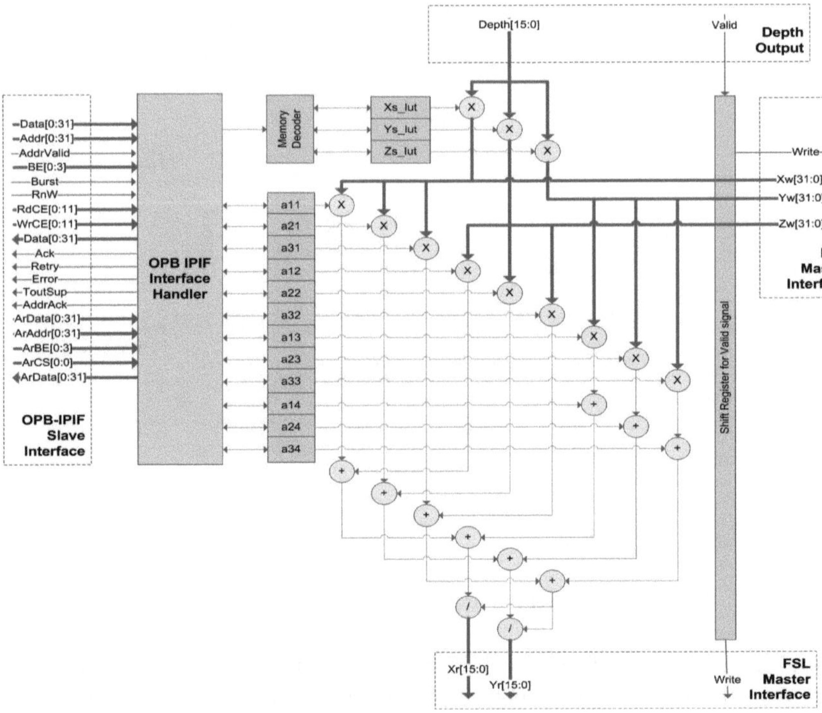

Fig. 8. Registration hardware architecture

5 Experiments

5.1 Error Analysis

Experimental error can be evaluated by applying the registration process over the circular marks extracted from calibration pattern images. The absolute error is given by the difference between the registered points to the correspondent circle centre in the second image. Once these points lie between the pixels, the used depth is calculated through bilinear interpolation of its neighbors. The error for 96 points at x and y coordinates is shown in Fig. 9. The outliers are caused by bad depth data from the camera, when not enough light is absorbed by the sensor. Fig. 10 shows the depth of each point. Each pattern image provides 6 points, which should remain at close depth values, making easy to observe invalid values.

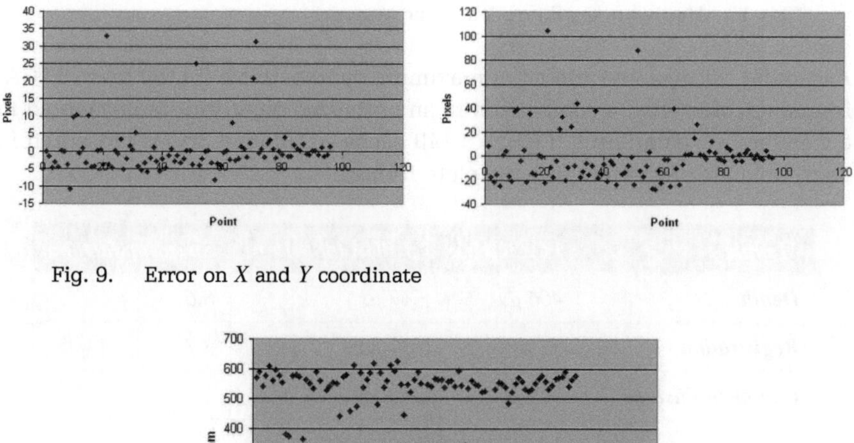

Fig. 9. Error on X and Y coordinate

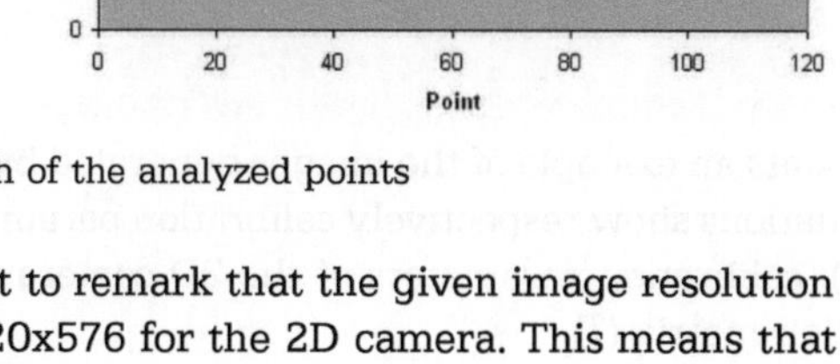

Fig. 10. Depth of the analyzed points

It is also important to remark that the given image resolution is 64x16 for the 3D camera and 720x576 for the 2D camera. This means that one pixel of the low resolution image is represented by 11.25 pixel width and 36 pixel height in the high resolution image.

5.2 Performance

Developed in VHDL and synthesized by the Xilinx ISE 7.1 [Xil06], the fusion core presents the following characteristics:

	Depth		*World*		*Projection*	
Slices	*2336*	*16%*	*132*	*<1%*	*4848*	*33%*
BlockRAMs	*3*	*3%*	*6*	*6%*	-	-
Mult 18x18	*6*	*6%*	*6*	*6%*	*36*	*37%*
Max. Frequency (MHz)	*88*		*280*		*127*	
Latency (clock cycles)	*63*		*1*		*37*	

Tab. 1. Hardware configuration in the system

Assuming the system running at maximum synthesizable (in the given FPGA) frequency (88 MHz), a comparison to an optimized algorithm implemented in a standard PC (Pentium 4 3 GHz - 1 MB cache - 1GB memory) is presented in next table for processing one complete frame.

Module	*PC (3GHz)*	*FPGA (88MHz)*	*Speed-up factor*
Depth	*406 μs*	*47 μs*	*8.6*
Registration	*37 μs*	*12 μs*	*3.1*
Complete Fusion	*443 μs*	*59 μs*	*7.5*

Tab. 2. System performance for dealing with one frame

5.3 Example

This section presents an example of the images generated by the system. First, the following equations show respectively calibration parameters matrix of the 3D camera (K_{3D}), calibration parameters of the 2D camera (K_{2D}) and the relative transformation matrix (T).

$$K_{3D} = \begin{bmatrix} 71.090520 & 0 & 39.849392 \\ 0 & 52.923175 & 8.663061 \\ 0 & 0 & 1 \end{bmatrix} \tag{18}$$

$$K_{2D} = \begin{bmatrix} 925.760849 & 0 & 373.528280 \\ 0 & 1013.488068 & 273.059471 \\ 0 & 0 & 1 \end{bmatrix} \tag{19}$$

$$T = \begin{bmatrix} 0.991048 & -0.046965 & 0.124974 & 3.997788 \\ 0.039669 & 0.997394 & 0.060243 & 110.319310 \\ -0.127477 & -0.054746 & 0.990329 & 147.667955 \end{bmatrix} \tag{20}$$

Fig. 11, and 12 show respectively the original and registered images of a person.

Fig. 11. Original images

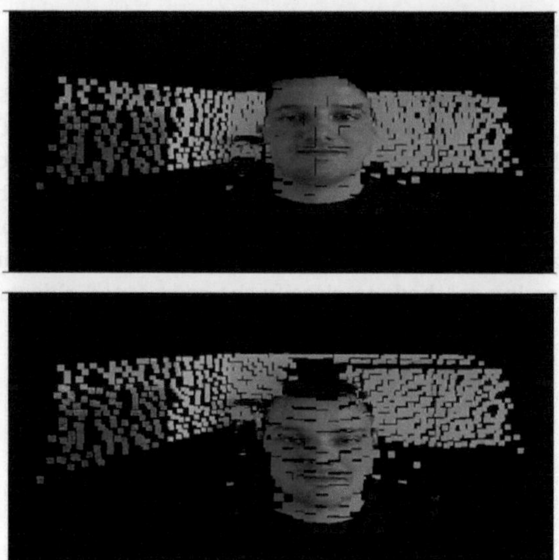

Fig. 12. Registered images

The 3D representation uses squares to show each points of the low resolution camera in 3D, while the registered 2D image is mapped as textures on the points. Observe that, on the object borders where the light gets reflected from multiple objects, the depth values of the points get smoothed. As shown in the results, the alignment error is lower than one pixel width, which can be also seen clearly at the horizontal and vertical borders.

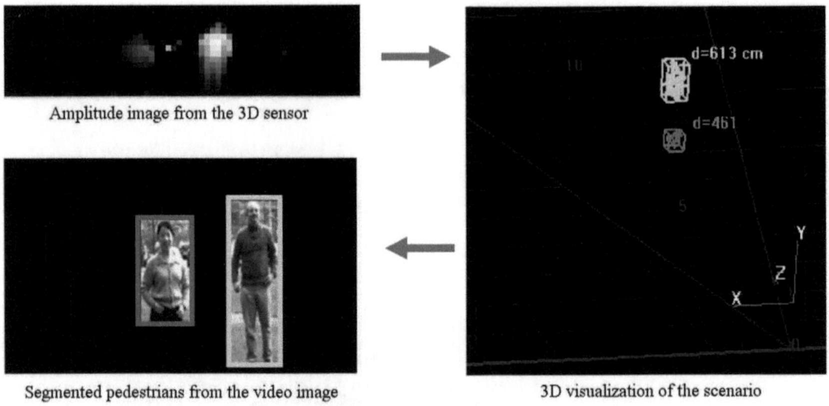

Fig. 13. Scenario for pedestrian tracking and classification

5.4 Further Related Experiment Scenario

Moreover, we have also built another experiment scenario closely related to this work. See Fig. 13. The experiment aims to track and classify pedestrians, cyclists and vehicles in traffic scenes. Currently the tracking and classification algorithms are based on the analysis of the range images grabbed from the 3D sensor, and the video images from the 2D sensor are fused for favorable 2D and 3D visualization. An interesting future research is to combine the 2D and 3D image-processing methods to classify the segmented objects.

6 Conclusions and Outlook

An efficient method for the fusion of 3D and 2D sensor was presented in this paper. The results show that this approach is already precise enough for applications that do not require critical information on object borders. Accuracy on object borders could be increased by local analysis using the given fusion result as a first guess. Moreover, auto-calibration methods might be further applied to constantly improve the alignment. The actual solution is compact, which is convenient to be used in embedded systems. Applications like tracking can be improved with the increasing information given by the fusion. The described method is not limited to time of flight and standard video sensors. Other sensors like thermal sensors (e.g. used in night vision systems) and laser scanning systems could be used in equivalent manner.

References

[1] Celoxica: RC203 Website, 2006, http://www.celoxica.com/products/rc203.

[2] R. Hartley, A. Zisserman: Multiple View Geometry In Computer Vision, Cambridge, 2000.

[3] X. Luan: Experimental Investigation of Photonic Mixer Device and Development of TOF 3D Ranging Systems Based on PMD Technology, PhD thesis, University Siegen, 2001.

[4] L. Tao, H. Ngo, M. Zhang, A. Livingston, V. Asari: A Multi-sensor Image Fusion and Enhancement System for Assisting Drivers in Poor Lighting Conditions, in AIPR'05: Proceedings of the 34th Applied Imagery and Pattern Recognition Workshop (AIPR'05), IEEE Computer Society, Washington, DC, USA, 2005, S. 106–113.

[5] Xilinx: Website, 2006, http://www.xilinx.com.

[6] Z. Xu, R. Schwarte, H. Heinol, B. Buxbaum, T. Ringbeck: Smart Pixel – Photonic Mixer Device (PMD): New System Concept of a 3D-imaging Camera-on-a-Chip, in International Conference on Mechatronics and Machine Vision in Practice, Nanjing, China, 1998, S. 259–264.

[7] Z. Zhang: A Flexible New Technique for Camera Calibration, IEEE Transactions on Pattern Analysis and Machine Intelligence, Bd. 22, Nr. 11, 2000, S. 1330–1334.

A. Linarth, B. Liu, O. Jesorsky, R. Kompe
Driving Assistance and Sensor Information
Elektrobit Automotive Software
Frauenweiherstr. 14, 91058 Erlangen
Germany
andre.linarth@elektrobit.com
bing.liu@elektrobit.com
oliver.jesorsky@elektrobit.com
ralf.kompe@elektrobit.com

J. Penne
Department of Pattern Recognition
Friedrich-Alexander University, Erlangen-Nuremberg
Martensstr. 3, 91058 Erlangen
Germany
penne@informatik.uni-erlangen.de

Keywords: 2D-3D Fusion, rigid registration, FPGA, calibration

Pedestrian Protection Systems using Cooperative Sensor Technology

R. Raßhofer, D. Schwarz, BMW Group Research and Technology
E. Biebl, C. Morhart, Technische Universität München
O. Scherf, S. Zecha, Siemens Restraint Systems GmbH
R. Grünert, BARTEC GmbH
H. Frühauf, Fraunhofer Institute for Integrated Circuits IIS

Abstract

Conventional pedestrian protection systems utilizing contact- or non-cooperative perception sensors are limited by uncertainties in target classification. Moreover, these systems offer no benefit in case of fully or partially hidden pedestrians like e.g. children hidden by cars. In this paper, we present a novel approach using cooperative sensor technology to overcome these drawbacks. Within the research project AMULETT, we are developing a cooperative sensor system to clearly classify and locate vulnerable road users. The system warns the driver in the early phase of a possible collision or is triggering collision mitigation strategies in case of unavoidable contact.

1 Introduction

The European Commission's Transport White Paper (2001) proposed the ambitious target of halving the number of road fatalities by 2010 [3]. Recently, great efforts have been made to improve pedestrian safety while various active safety systems based on perception sensor technologies like radar, lidar or cameras (near and far infrared, mono and stereo vision) have been subject of extensive research [4, 8].

Lidar-, radar- or vision-based approaches offer highly accurate positioning information. Classification of objects based on these sensors' data can be difficult, particularly in scenarios where potential collision opponents like pedestrians are partially or fully hidden to the sensor shortly before an accident happens. These technologies are not able to detect objects without a line-of-sight contact. In-depth accident studies showed that in about 40% of all fatal pedestrian accidents the collision opponent was hidden to the driver until shortly before the accident happened. Object classification for advanced high quality driver assistance systems can be made more reliable if present sensor technologies are supported by cooperate sensor technology.

Cooperative sensor systems are using reactive targets, being able to identify themselves after being pinged by an RF interrogation pulse. From the received echo of the target, its position and type is determined. Diffraction of waves provides contact even if the tag is optically hidden to the receiver. Cooperative sensor principles have been extensively used by e.g. air traffic control ("secondary surveillance radar") for over 60 years now and proved to provide an extremely high level of reliability [5].

1.1 Related Work

L. Andreone et al. [1] propose cooperative systems for vulnerable road users based on a combination of a localisation unit for active tags and a near infra-red CMOS-camera. The positioning of the tag by triangulation is described in [7]. A cooperative sensor system using triangulation based on the detected DOAs at two subsequent locations along the vehicle moving direction is presented by J. Shi et al. [6].

In this paper, the cooperative sensor system is consisting of a transmitter operating the interrogation pulse, a smart antenna array determining the wave vector of the incident target response and time-of-flight measuring equipment determining the distance of the target. Moreover, decoding the modulation of the target's echo it is classified as vulnerable road user.

1.2 Overview

This publication focuses the sensor concept for a cooperative sensor system. The two main parts of the sensor concept are presented in chapter 2. For the active part the measurement of the distance (2.2.1) and the angle of arrival (2.2.2) are described in more detail. Furthermore the suitability of different ISM bands for this application is discussed (2.2.3). Results of in-depth accident studies underline the need for cooperative sensor systems in chapter 3. Finally, conclusions are made and an outlook on further work in the project is given.

2 Sensor Concept

A central part of the system is the planar antenna array attached to the front window of the vehicle. With this array the presence of an active tag and its position can be determined. For the passive part of the transponder a tag reader is integrated into the car attached to the car body used as a receiver.

1.1 Passive Short-Range Tag

The passive tag offers protection in situations where the pedestrian is located close to the car. A typical situation is a parking-lot where children walk behind a car driving in reverse. The car builds a safety zone of a few meters around itself and even around attached trailers.

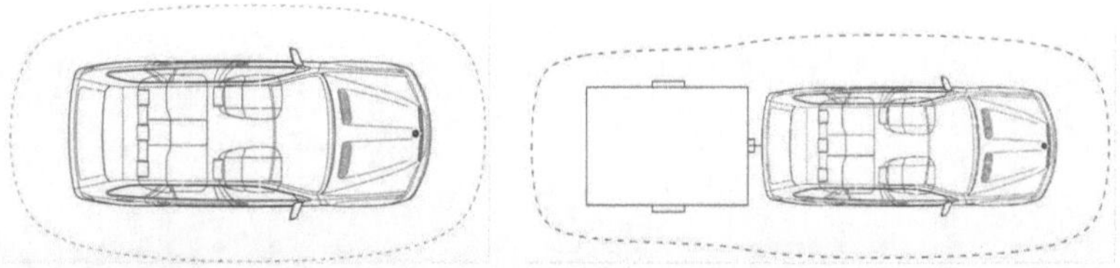

Fig. 1. Near-field of vehicle and vehicle with attached trailer

Tags located within this zone can be contacted by near field communication [2] to give a well-timed warning to the driver, indicating the presence of vulnerable road users or other obstacles dangerous for the car.

Measurements showed that within the safety zone, a tag is detected in every possible position. The presence of a tag can be detected even if it is located under the vehicle. This ability even enables the cooperative sensor system to protect children playing or sitting under a truck or bus as there are no blind areas.

The driver gets optical and acoustical feedback and can acknowledge the perception of a warning caused by a certain tag. In doing so he shuts down the warning for a tag, if it is a located in his field of vision and is recognized by him for his further driving actions.

This part of the cooperative sensor system offers assistance to the driver especially when maneuvering or at low velocities.

1.2 Active Long-Range Tag

For the protection of vulnerable road users in urban and extra-urban traffic, longer detection ranges are necessary. Furthermore, the ability to detect and locate fully or partially hidden tags provides a new feature in vehicle sensor and perception systems. In these cases, the tag's response is received by a smart antenna array, measuring its azimuth angle and distance along with its (anonym) identity used for classification and tag separation. Investigations showed that a line-of-sight between the tag and the antenna array is not required to detect the presence of the tag since diffraction of electromagnetic waves provides contact between the transponder and the receiver.

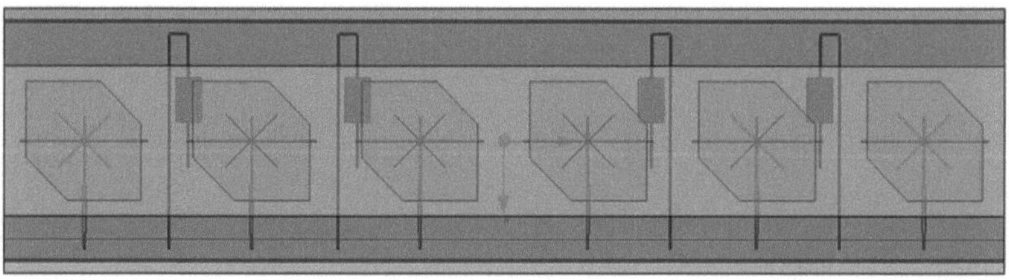

Fig. 2. Planar antenna array used for localization, with a size of less than 0.5 m it can be integrated at the top of the car's front window

1.2.1 Measurement of the Distance

The tag's distance can be calculated by a time-of-flight measurement, determining the round trip delay of the interrogation pulse. Furthermore the change of the measured azimuth angle can be investigated knowing the vehicle's velocity to achieve more accurate positioning [6]. Time-of-flight measurement is done by means of a sub-carrier technique. The car sends an interrogation sequence to the tag. The tag measures the phase difference between the interrogation sequence and its internal time-base and responds after a certain time-delay by sending information about this time delay, its ID-code and the measured relative phase. Since each tag uses a different time-delay, the responses of several tags being simultaneously interrogated are separated in time enhancing the precision of the measurement of the distance and the direction-of-arrival. After a certain period the tag sends a second response that is used for calibration of the clock.

1.2.2 Measurement of the Direction of Arrival

Using a miniaturized planar antenna array, an estimation of the direction of arrival (DoA) can be obtained. The direction of the tag's signal is calculated via algorithms like ESPRIT (Estimation of Signal Parameters via Rotational Invariant Techniques). Further shrink of package size might be obtained by the use of micro system technology. With this information the relative position of the pedestrian can be calculated and the movement can be tracked by different processing units.

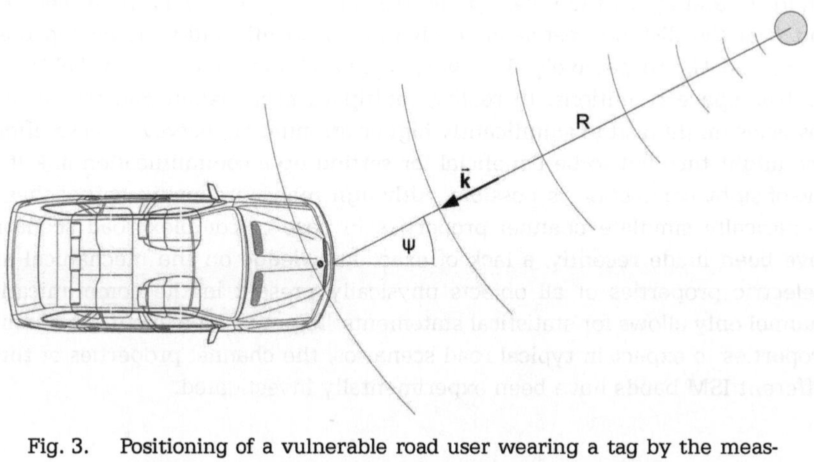

Fig. 3. Positioning of a vulnerable road user wearing a tag by the measured distance and DoA (Direction of Arrival), R: Range of tag, ψ: Incident angle of wave vector, k: wave vector of tag's response

In first measurements the tag's azimuth angle could be determined with an accuracy of about 1 degree in different distances in case of a line-of-sight connection. The accuracy of the angular measurement is expected to decrease without a line-of-sight connection. However, still good measurement quality is expected, depending on SNR and multipath scenario. These effects have to be investigated in future experiments, planed in a catalogue covering the most common urban and extra-urban scenarios.

1.2.3 ISM Bands

In a first approach, the system should use ISM bands to ease international licensing. As each ISM band offers different channel properties, in-depth analysis of the pros and cons of each band is necessary.

If we assume a line-of-sight connection, the channel attenuation A_{ch} is determined by the Friis transmission equation (1)

$$A_{ch} = \frac{P_r}{P_t} = \left(\frac{\lambda}{4\pi R}\right)^2 G_{0t} G_{0r} \tag{1}$$

where P_r and P_t are the received and transmitted power, λ is the wavelength and R is the distance between receive and transmit antennas having gains of G_{0t} and G_{0r} respectively. This very simple channel model only holds true for free space conditions. In reality, multipath propagation and presence of obstacles might lead to significantly higher attenuation, however these effects also might turn out to be beneficial for setting up a communication link if no line-of-sight connection is possible. Although many promising approaches to numerically simulate channel properties in case of complex road scenarios have been made recently, a lack of exact knowledge on the mechanical and dielectric properties of all objects physically present in the communication channel only allows for statistical statements. To get an idea about the channel properties to expect in typical road scenarios, the channel properties of three different ISM bands have been experimentally investigated.

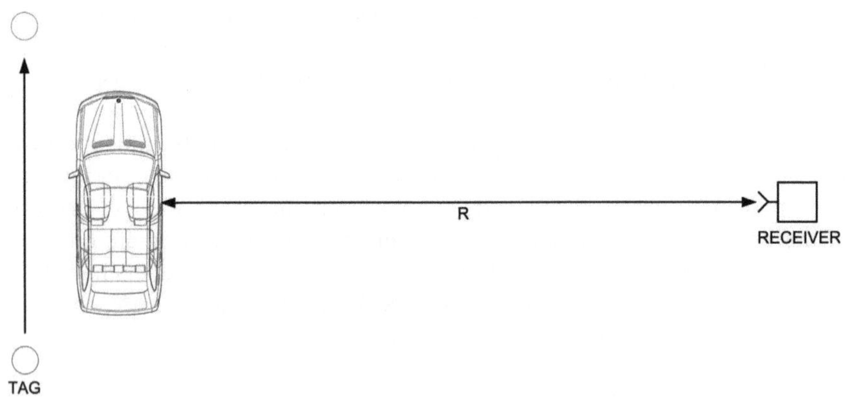

Fig. 4. Measurement setup used to determine the channel properties of different ISM bandsin case of a pedestrian hidden by a car

For pedestrian protection, communication with a tag being optically hidden behind a car represents one of the most challenging scenarios. Measurement of channel properties in this case has been done by using a setup shown in Fig. 4. The tag was carried behind a car in various heights and the received signal strengths were measured.

In Fig. 5 (a) the received field strengths for a tag communicating on 868 MHz (blue) and a tag communicating on 2.4 GHz (red) are compared at different positions behind the car. The quasi-optical character of higher frequencies like 2.4 GHz results in better transmission through the windows of a car as can be seen in Fig. 5 (b).

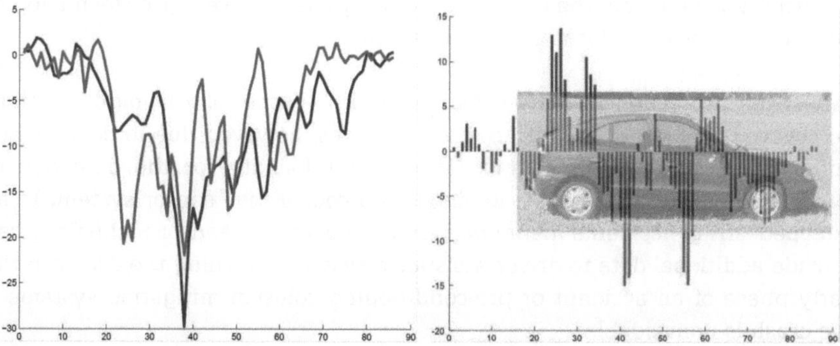

(a) (b)

Fig. 5. (a) Received signal strength of tags transmitting on 868 MHz (blue) and 2.4 GHz (red), (b) difference between received signal strength in dB, blue: amplitude of 868 MHz signal is higher, red: amplitude of 2.4 GHz signal is higher

Here the difference of the received signal strengths of a tag carried in a height of 80 cm in dB is shown along the car. If there is a line-of-sight like through the car's windows the signal strength of 2.4 GHz is higher at the receiver. At the other spots like behind the a-pillar the signals of the 868 MHz tag is dominant because of its better diffraction properties.

These results lead to the conclusion to use two different frequencies in the tag. With a signal of 868 MHz the presence of a tag can be determined more reliable especially in cases where the tag is hidden. The transmission of a signal at 2.4 GHz is used for the positioning of the tag. One advantage of the usage of 2.4 GHz for positioning is that antenna spacing in the array can be reduced compared to 868 or 434 Mhz as it relates to the wavelength.

3 Accident Studies

In-depth accident studies showed that over 40% of all fatal pedestrian accidents in Germany were induced by late recognition of the dangerous situation by the driver. Most critical are situations where a pedestrian is either hidden, stands behind the vehicle e.g. on a parking-lot or walks along an extra-urban road with no illumination at night. In scenes where pedestrians are hidden by obstacles and step onto the road abruptly, only cooperative sensor technologies provide a chance to inform the driver in time.

Pedestrian protection in extra-urban scenarios – especially at night – might be covered by near or far infrared technologies, however, due to curvy roads and missing line-of-sight contact, increased reliability for the detection of pedestrian or bicyclist will be obtained by a cooperative sensor system. In all fatal pedestrian accidents mentioned above, cooperative sensor technology can provide additional data to driver assistance systems, warning the driver in the early phase of an accident or pre-conditioning collision mitigation systems if the crash is unavoidable.

Based on the accident studies, a catalogue of different scenarios has been developed covering the most critical situations. These scenarios will be used to evaluate the proposed system in respect to collision avoidance and collision mitigation. The maximum detection range, the ability to separate different tags from each other and the accuracy of hidden tag localization will be evaluated in these different experimental scenarios.

4 Conclusion

This paper proposed a novel approach for vulnerable road user protection using cooperative sensor technology. The system will be developed by the AMULETT project consortium. It is based on RF-communication technologies between a low power transponder attached to vulnerable road user and a receiver unit built into a vehicle, capable to precisely localize the transponder. The system does not suffer from the limitations of conventional sensor-based approaches since it provides reliable target classification and accurate localization even if no line-of-sight connection is possible or if bad weather conditions prevail.

Cooperative sensor technology used for pedestrian protection is another promising mass market application for modern micro system technology in the automotive environment.

5 Further Work

Further work includes the development of tracking algorithms using cooperate sensor data. Based on the results of the tracking, warning algorithms or active safety concepts can be implemented. One of the main challenges of the project is to be able to handle different urban and extra-urban scenarios that cover a huge number of different types of accidents involving vulnerable road users. Extensive system tests will be conducted ensuring a proper evaluation of the sensor system in the mentioned scenarios. Therefore the described components will be integrated into a test vehicle used to obtain realistic data for system evaluation.

With a further shrink of the package size the transponder could even be integrated into mobile devices like mobile phones or into clothes as wearable solution.

6 Acknowledgment

The cooperative sensor system presented in this publication is part of the main efforts in the AMULETT-project which is co-funded by the VDI/VDE Innovation + Technik GmbH. The project aims to avoid accidents involving vulnerable road users such as pedestrians and bicyclists by well-timed warnings and active safety concepts.

References

[1] Luisa Andreone, Andrea Guarise, Francesco Lilli, Dariu M. Gavrila and Marco Pieve. Cooperative Systems for vulnerable road users: The Concept Of The WATCH-OVER Project, ITS 2006

[2] Near Field Communication White Paper. ECMA international, 2004

[3] European transport policy for 2010: time to decide, WHITE PAPER, 2001

[4] D. M. Gavrila, J. Giebel and S. Munder. Vision-based Pedestrian Detection: the Protector System, Proc. Of the IEEE Intelligent Vehicles Symposium, Parma, 2004
[5] Werner Mansfeld, Funkortungs- und Funknavigationsanlagen, 1994
[6] Jianliang Shi, Hans Christian Mueller and Michael Marx. Pedestrian Detection and Localization Using Antenna Array and Sequential Triangulation, ITS 2006
[7] Prof. Dr.-Ing. Dipl.-Ing. Dipl- Wirt.-Ing. Axel Sikora. Localization with Short-Range Wireless Networks, Wireless Congress, 2006
[8] L. Walchshäusl, R. Lindl, K. Vogel, T. Taschke. Detection of Road Users in Fused Sensor Data Streams for Collision Mitigation, AMAA 2006

R. Raßhofer, D. Schwarz
BMW Group Research and Technology
Hanauerstr. 46
80992 Munich
Germany
Ralph.Rasshofer@bmw.de
Daniel.Schwarz@bmw.de

E. Biebl, C. Morhart
Technische Universität München
Fachgebiet Höchstfrequenztechnik
Theresienstr. 90 / N8
80333 Munich
Germany
Biebl@tum.de
Morhart@tum.de

O. Scherf, S. Zecha
Siemens Restraint Systems GmbH
Carl-Zeiss-Str. 9
63755 Alzenau
Germany
Oliver.Scherf@siemens.com
Stephan.Zecha@siemens.com

R. Grünert
BARTEC GmbH
Schulstr. 30
94239 Gotteszell
Germany
Rudolf.Gruenert@go.bartec.de

H. Frühauf
Fraunhofer Institute for Integrated Circuits IIS
Am Wolfsmantel 33
91058 Erlangen
Germany
Holm.Fruehauf@iis.fraunhofer.de

Keywords: cooperative sensors, pedestrian detection

Powertrain

Misfire Detection System based on the Measure of Crankshaft Angular Velocity

F. Lo Bue, A. Di Stefano, C. Giaconia, E. Pipitone
Università degli Studi di Palermo

Abstract

Misfire detection systems are becoming increasingly important in automotive market due to recent environmental issues (Euro rules). An early misfire diagnosis also allows to prevent damages to the exhaust emission system and consequent costs for the user. Today few low cost methods exists in order to precisely detect single misfires in real time, the majority in fact require the use of expensive sensors (e.g. pressure sensors) or dedicated circuits (e.g. ionization current sensing). This work describes a method and electronic system capable of detecting misfires with good accuracy, using parameters such as the speed sensor signal, already available in commercial engines. The proposed method is exclusively based on the real time analysis of the crankshaft angular velocity and its variations. This is possible since each misfire event generates an abrupt perturbation of the crankshaft angular velocity, that can be detected using an appropriate signal processing algorithm.

1 Introduction: The Misfire Detection Issue

Today's automotive engines must meet both fuel economy requirements and pollutant emissions regulations, which become more and more stringent. Misfires detection has become a very important task in pollutant emission control, since a lack of combustion causes unburned gasoline to enter the exhaust pipe, producing high and sometimes even unacceptable level of unburned hydrocarbons; moreover gasoline flowing inside the exhaust pipe can burn inside the catalyst oxidation ambient, causing high temperature rise which in turn can seriously damage the catalyst substrate, further weakening the exhaust gas after treatment system. For this reason On Board Diagnostic (OBD) regulations require the engine's Electronic Control Unit (ECU) to be able to detect misfires occurrences in order to warn the driver on the required technical assistance. Researchers all over the world therefore focused on setting up and testing systems and techniques for misfires recognition, which are mainly based on the analysis of:

- in-cylinder pressure
- ionization current or Breakdown voltage
- crankshaft angular speed
- temperature and Oxygen concentration

In-cylinder pressure is obviously the most reliable quantity for misfire detection purpose, since a lack of combustion will make the cylinder to run "motored", resulting in a negative Indicated Mean Effective Pressure (IMEP). However the endowment of each engine cylinder with a combustion chamber pressure sensor is not economically feasible, hence the necessity to find an alternative way to recognize misfires. The ionization current, flowing between the spark electrodes after combustion has started, is strictly related both to pressure evolution and flame propagation, since it depends on the ions concentration in the electrodes gap: its analysis therefore allow to surely detect misfire occurring, making this method one of the best alternative to in-cylinder pressure measurement. However it requires some electric circuitry modifications and signal amplification (or high voltage supply) [1-3]. A little different approach has also been tried out by measuring the spark gap breakdown voltage, but some more effort will be necessary in order to become competitive with the ionization current method.

A very easy-to-measure quantity for misfire detection purpose is the crankshaft angular speed: the lack of combustion in fact causes a sensible perturbation in engine speed progress, which can be detected by means of the common crankshaft position sensor usually equipping automotive engines. This methods is actually one of the most investigated, since nor additional sensors are needed neither any modification to electric circuitry is necessary [4-6]. The most followed approaches are two: in the first one the instantaneous rotational speed may be analysed in order to reconstruct indicated torque; the second instead performs statistical evaluations over a certain number of cycles. However, complex signal processing may be required in the latter case since the perturbation caused on the engine rotational speed by a single misfire is not so evident under certain operative conditions (low load and/or high speed).

Exhaust gas temperature and oxygen concentration are also easy-to-measure quantities which undergo a sensible variation as consequence of misfires, but the resulting frequency response is to low and, unlike the other mentioned methods, the measure of these two quantities cannot give precise information on which cylinder is misfiring [7]. The measure of these quantities may be used in conjunction with the other methods for a better understanding of the misfires causes: a fault in the injection system in fact would cause high oxygen concentration on the exhaust gas, while a fault in the ignition system may cause also a temperature rise.

2 Design and Electronic Implementation

This paper describes the design and implementation of an electronic system for misfire detection based on the measure of the crankshaft angular speed. This is mainly due to its large availability within almost all the vehicles present in the nowadays market and because its measurement is the most economic and less invasive. The misfire detection system main goal is to get a real time monitoring of the engine, in order to detect one or more misfire events. Besides this objective the electronic system must be functionally portable in order to allow a simple integration within digital control systems already present in the automotive marketplace. These constraints ended up to the design and implementation of algorithmically simple solutions based on the adoption of a low-end microprocessor and the use of simple arithmetic operations. An 8-bit microcontroller with a relatively low processing capability was then chosen. In particular it doesn't implement any sophisticated Digital Signal Processing (DSP) function and, as a consequence, a 16 bit fixed-point signed arithmetic was adopted throughout all the calculations carried out to the purpose. The algorithm is composed by the following steps:

- instantaneous crankshaft angular speed measurement;
- data filtering;
- smart comparison between present and past acquired data;
- generation of suitable parameters strictly related with the misfire probability;
- run-time evaluation of misfire events within one engine cycle.

The measure of the instantaneous angular speed is of course the most important sensing measure of this method since its precision is a direct evaluation of the overall resolution of the system. The method in fact tries to extract results on possible misfire events from the crankshaft speed variations. To this purpose the square wave signal coming out from a standard optical encoder, rigidly connected to the engine crankshaft, was used as a trigger for a 16 bit deep timer, internal to the microcontroller, running at the clock frequency of 18.432 MHz. This frequency guarantees a very precise measure of timing intervals. In particular it has been decided to trigger the internal timer every 4 crank angle degrees which in turn corresponds to about 111 μs in the case of an engine speed of 6000 rpm. By comparing it with the inverse of the clock frequency (54.25 ns) it's clear that timer counts of about 10^3 are assured, leading to a minimum precision of at least three orders of magnitude. If the engine is equipped with a standard position and speed measuring system, based on the use of an inductive sensor and a 60 teeth pulse ring, the trigger will happen every 6 crank angle degrees thus causing only the needs of minor changes on the above explained choices.

Unfortunately the acquired data are deeply affected by several noise sources mainly due to mechanical oscillations and strong electromagnetic interferences generated by the engine. Fig. 1 shows the normalized FFT (Fast Fourier Transform) of the acquired data at different engine speeds: a clear signal is visible at frequency F_0, corresponding to the gas expansion within the cylinders. Others more or less significant contributes are present most of which are generally not related to the misfire phenomena hence they could potentially be considered as an overlapped noise. Furthermore the behaviour of frequency components quite dramatically change at different engine speeds.

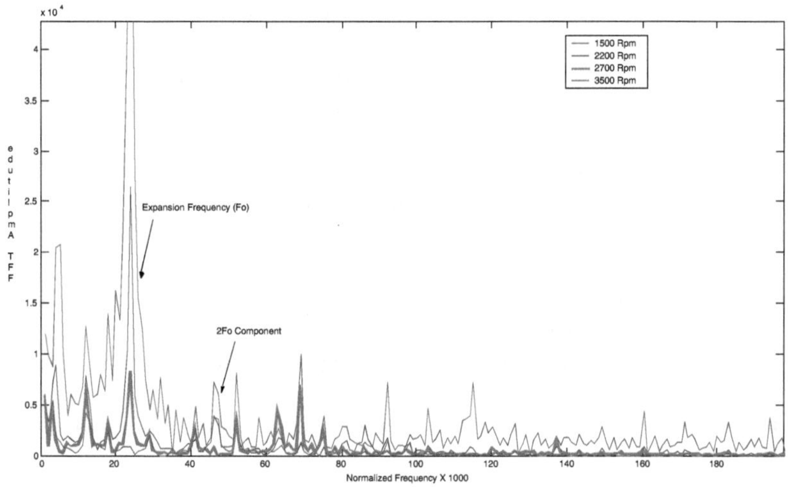

Fig. 1. FFT comparison of timing fluctuations at different engine rotation speeds

Now these spectra may in some sense be considered as the fingertips of a particular engine, hence data filtering must be tailored on every single vehicle through a learning phase capable to customize the filter characteristics. In order to reduce noise and isolate frequency components of interest, the samples are processed using a digital Infinite Impulse Response (IIR) filter and the obtained signal shows a good regularity in normal condition (i.e. signals related to two subsequent engine cycles are very similar), while it exhibits significant changes when a misfire occurs. This property has been exploited in order to detect misfire events, by evaluating the absolute difference of the signals related to two subsequent cycles. To find out the main features of the digital IIR filter it has to be pointed out that speed deviations can be ascribed to the inertia of the reciprocating mass hence it is possible to design the shape of the filtering action by pursuing the main goal to emphasize the normal engine cycle; this is obtained by properly depressing vibrations effects and other phenomena not strictly related to pistons' acceleration. In fact the indicated torque C_i, that is the gas pressure torque, is related to the crankshaft speed and acceleration by the following equation:

$$C_I - C_F - C_B = (I_C + I_R) \cdot \frac{d\omega}{dt} + \frac{1}{2}\omega^2 \cdot \frac{dI_R}{d\vartheta} \qquad (1)$$

where C_F and C_B are the friction torque and the braking torque, I_C is the moment of inertia of the crankshaft with respect to the crankshaft axis, I_R represents the equivalent moment of inertia of the reciprocating masses and finally ω represents the angular velocity.

As it can be seen, the net torque equals the sum of two terms: the first is linearly related to the crankshaft acceleration, whose variation may give a measure of the torque The second one instead is linearly related to the piston and connecting rod inertia variation, and its absolute value is weighted by the squared angular speeds, thus its effect is negligible at low engine speed and becomes increasingly important with growing speeds. It follows that two factors affect the crankshaft acceleration within one engine cycle: one due to the gas expansion with a fundamental frequency (F_0) related to the angular acceleration an the other one caused by the variations of the moment of inertia of the reciprocating masses; the latter has the same frequency of the first but different phase (I_R reaches a maximum approximately 90 degrees after top dead centre), thus a characteristic frequency at $2F_0$ is expected in normal conditions (no misfires detected).

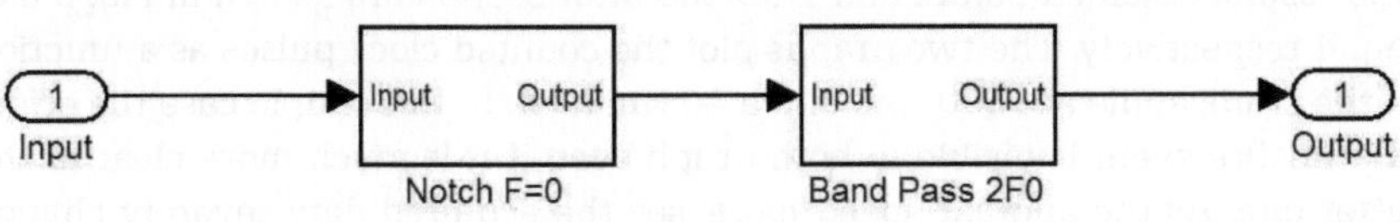

Fig. 2. Diagram of the IIR digital filters blocks

In order to emphasize the above mentioned effect a band pass filter, centred in $2F_0$, will amplify the system response at high engine speed, where the former contribute strongly decays (see 2700 rpm graph of Fig. 1) or almost disappears (see 3500 rpm graph of Fig. 1). Furthermore another notch filter at $F=0$ is also used with the purpose of obtaining the advantages to get rid of the meaningless continuous frequency components that limit the digital dynamic of the IIR filter and the low frequency oscillations due to mechanical resonance.

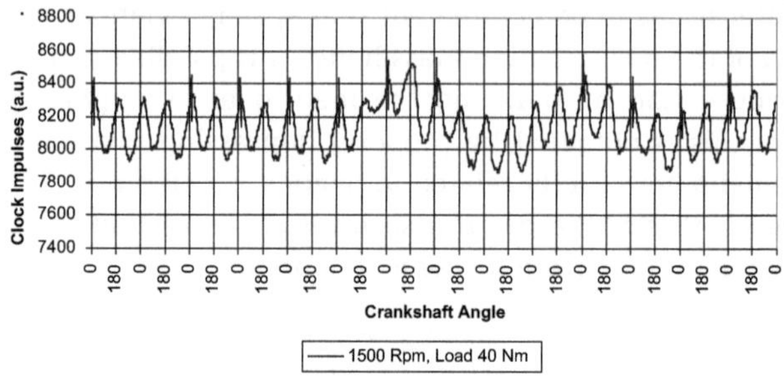

Fig. 3. Unfiltered speed fluctuations (1500 rpm and 40 Nm load)

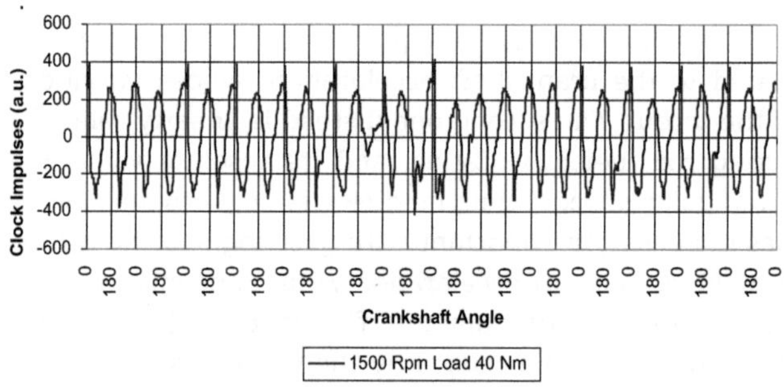

Fig. 4. Filtered speed fluctuations (1500 rpm and 40 Nm load)

The results obtained before and after the filter section are shown in Fig. 3 and Fig. 4 respectively. The two graphs plot the counted clock pulses as a function of the crank angle at 1500 rpm and a 40 Nm load. In this simple case the effect of a misfire event is visible in both graph even if it is much more clear in the latter one. As the angular speed increases the acquired data severely change their shape and the unfiltered data progressively become incomprehensible as the next couple of Fig. [5, 6] show.

In particular 4 engine cycles without misfire occurrences are represented: the filtered samples confirm the presence of a periodical behaviour on every engine cycle, while the unfiltered values seem to be quite randomly scattered and consequently they do not permit any misfire detection based on a regular signal under normal circumstances.

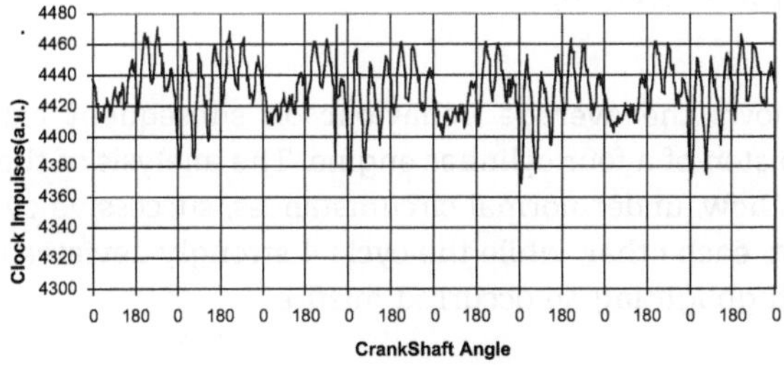

Fig. 5. Unfiltered speed fluctuations (2700 rpm)

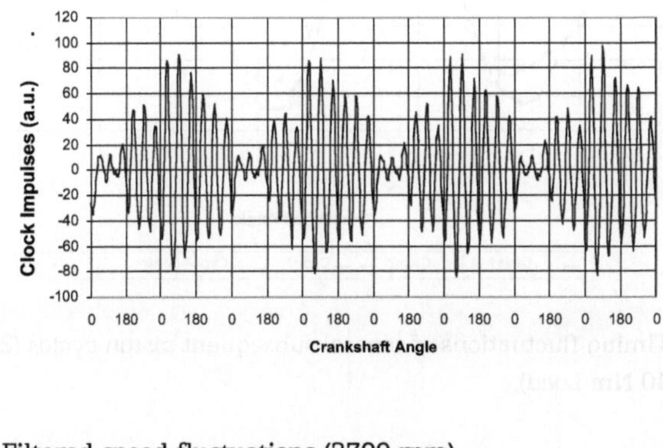

Fig. 6. Filtered speed fluctuations (2700 rpm)

The implementation of the IIR filters with the adoption of a 16-bit fixed point arithmetic imposes particular care in the choice of the approximated coefficients of the filters and in a skilful use of binary maths by massively exploiting, whether possible, the use of power of 2 numbers in order to minimize the computational load for the selected microcontroller while keeping stable the filters behaviour, minimizing deviations from the ideal digital filters, and finally keeping low truncations and rounding-off errors.

The filtered data are then passed to another firmware-implemented block of the digital system whose function is to continuously compare running data and to generate or update the reference parameters for the evaluation of a possible misfire event. First of all the temporal distances between angular speed data from every single piston cycle (i.e. 180 crank angle degrees) is calculated by using the last arrived 45 samples. They are compared with the ones read in the preceding piston cycle and the mean distance between these two sets of filtered speed data is updated with the following algorithm:

$$D(n)=\sum_{i=0}^{44}\left|y_i(n)-y_i(n-1)\right| \tag{2}$$

The Fig. 7 shows the average behaviour on subsequent expansion cycles referred to a piston of a four cylinder engine. The analysis of these experimental data shows how, under normal circumstances, successive cycles have little deviations from each other, while the cycle 4 strongly deviates from the mean curves hence highlighting an occurred misfire.

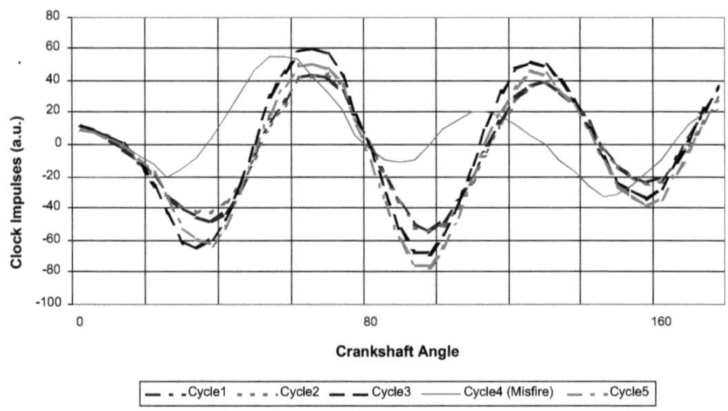

Fig. 7. Timing fluctuations of several subsequent piston cycles (2500 rpm, 40 Nm Load)

In order to compare piston cycles between each other, a moving average (*m*) of the last two *D(n)* values is calculated and then multiplied for an integer value *k* (between 1 and 3), experimentally evaluated and related to the engine speed.

$$M(n)=k\cdot m(n)=k\cdot\frac{m(n-1)+D(n)}{2} \tag{3}$$

This *M(n)* is the main parameter kept in account as a reference to evaluate misfires. More precisely if a value of *D(n)* exceeds the above mentioned average value (*M*) then it gives a start to the assessment of a possible misfire event. Now the above explained threshold is not yet sufficient to discriminate normal cycles from misfires. In fact the difference signal is strongly affected by the operating conditions, primarily speed and load variations, and then the simple evaluation of a dynamic threshold exceeding usually leads to an intolerable number of false misfires detection. In order to further increase the detection accuracy, the digital system quantifies and keeps into account:

- how much the moving average (*M*) has been exceeded and
- how many times it occurred in an engine cycle.

This last statement could appear strange at first sight, but it has to be pointed out that a real misfire event usually has an effect on the inertia variations of immediately subsequent piston cycles of the other cylinders, within the same engine cycle. This effect is shown in the Fig. 8 that represents the $D(n)$ values measured for each piston cycle. It's clear that the misfire event causes a sudden increase of $D(n)$ and it happens at least for two piston expansion in the same engine cycle.

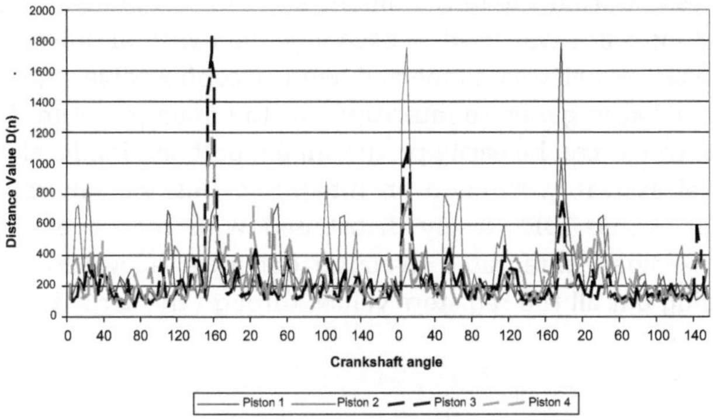

Fig. 8. D(n) values between two contiguous engine cycles for all the four pistons

These circumstances effectively increase the probability to have a correct misfire detection and then if this last probability goes beyond an experimentally evaluated threshold, a misfire detection alarm is finally registered in the ECU.

3 Mechanical and Electronic Experimental Setup

The method proposed in this paper, underwent many trials on the engine test bed of the Department of Mechanics at University of Palermo, endowed with a Fiat 1242cc four cylinders spark ignition engine connected to a Borghi&Saveri eddy current dynamometer. The engine was equipped with an AVL GU13X piezoelectric pressure transducer, installed in the combustion chamber by means of its spark plug adaptor ZC32, and the already mentioned 360 pulses per revolution optical encoder was used to clock the analogue acquisition of the engine data (torque, speed and in-cylinder pressure) with a resolution of 1 crank angle degree. The acquisition of all parameter was independently performed using a National Instrument 6062 DAQ card and the LabVIEW™ software.

As for the digital electronics system under test, a Zilog's development board equipped with a Z8F6403 microcontroller was used. The Z8F6403 has an 8 bit CPU, 4 Kbytes of general purpose data RAM and 64Kbytes of Flash program memory. The development board allows direct access of microcontroller's pins, support of serial and infrared interfaces and an handy 4 digit LED matrix intensively used during the development and the test phases. In particular a serial RS232 interface was used to exchange data with an host PC, while the optical encoder signal was connected with a configurable input pin of the Z8F6403, enabled to generate interrupts for the main program. Furthermore the microcontroller can be easily programmed, both at high level (C language) and low level assembly, through an Integrated Desktop Environment (Zilog Development Studio IDE), freely distributed with the development board. Finally an hardware integrated in-circuit debugger allows to read the CPU status bringing out all the registers values through the serial interface.

4 Tests and Results

In order to evaluate the performances of the misfire detection system and give evidence of its correct behaviour an autonomously misfire generation system is required. To this purpose the misfires were induced using a PC-based spark ignition control system already available at the Department of Mechanics, which employs two IGBTs (each one serving a couple of cylinders) to charge the primary coils, activated on their gates by means of TTL pulses sent by a National Instruments counter/timer board PCI 6602 managed under LabVIEW™: this system permitted to miss single ignition, simulating misfires due to ignition system fault. The same counter/timer board was also used for instantaneous rotational speed evaluation. The misfire detection method was then tested under different operative conditions of engine speed (from 1500 to 3000 rpm with steps of 500 rpm) and load (4 and 7 bar Brake Mean Effective Pressure). For each of them misfires have been randomly induced, verifying the response of the misfire detection system by means of the calculation of IMEP, which becomes negative when the cylinder is "motored", i.e. each time a misfire occurs. As an example a bar graph in a relatively low load condition is reported in Fig. 9. The results shown were carried out by accumulating the data coming from 4 tests, each one randomly generating 50 misfires. As it can be seen two different set of bars are collected at various speed regimes: the lighter ones represent the percentage of the total generated misfires events, correctly read by the detection system; the other ones instead testify the presence of false misfire detections, i.e. read by the detection system but not actually injected by the generation system. In the former case the performances hang considerably above 90% at all the different rotation regimes and this constitutes, in the authors' opinion, a firmly encouraging result. The latter case instead shows a very low level of false alarms till 2500 rpm while a net increase of them is experienced when the rotation speed grows up. At that regime in fact the absolute value of inertia fluctuations reach the height of the gas expansion contribute (Fig. 1) thus making more critic the misfire recognition and justifying the presence of the filtered amplification already discussed. However even in this difficult conditions, the difference between a true and a false misfire detection remains quite high.

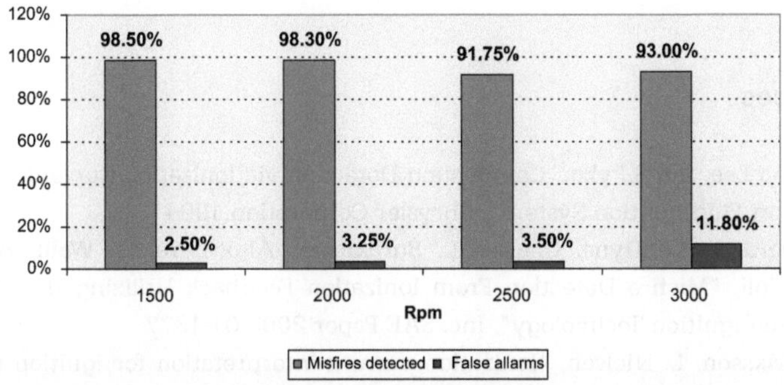

Fig. 9. Experimental results of collected of misfires and false alarms

5 Discussion and Conclusions

The proposed system, implemented with an 8 bit microcontroller and tested with a FIAT 1242cc 4 in-line cylinder 8 valve engine, was able to detect up to 98.5% of total misfires at lower speeds, and more than 90% at higher speeds. In few cases some false detections were obtained. However the proposed system and method generally achieved comparable or better results than conventional methods. Furthermore some final remarks are needed:

The firmware has been produced by coding all the routines in ANSI C, thus guaranteeing their great portability within a large range of microcontrollers; and if a CPU implements some DSP functions the software is ready to embed them within the designed solution thus speeding up all the calculations;

Since the detection system is based upon a comparative reading of subsequent piston cycles, all the phase shifts introduced by internal computations, regarding the digital filter section and the decision algorithm, do not affect the measures and the overall performances;

As for the k parameter, present in the moving average formula, and experimentally evaluated, it can be easily within the microcontroller flash memory after a calibration setup;

Finally the detection method is applied to all the engine pistons since a misfire event on one of them modifies the behaviour of the nearest piston cycles. That's consequently requires an over imposed engine cycle synchronization capable to distinguish the different piston cycles.

References

[1] Ansen Lee, Jan S.Pyko, "Combustion Detection via Ionization Current sensing for Coil on Plug Ignition System", Chrysler Corporation.1994

[2] Edward A. VanDyne, Charles L. Burckmyer, Alexandre M. Wahl, Alberto E. Funaioli, "Misfire Detection From Ionization Feedback Utilizing the Smartfire® Plasma Ignition Technology", Inc. SAE Paper 2000-01-1377

[3] L. Eriksson, L. Nielsen, "Ionization current interpretation for ignition control in internal combustion engine", Control Eng. Practice, Vol. 5 No. 8, 1997, Elsevier Science Ltd.

[4] I. Haskara, L. Mianzo, "Real-time cylinder pressure and indicated torque estimation via second order sliding modes", Proceedings of the American Control Conference, June 2001

[5] John J. Moskwa, Wenbo Wang, "A New Methodology Utilizing "Synthetic" Engine Variables - Theoretical and Experimental Results", Transactions of the ASME, Vol. 123, September 2001

[6] Jinseok Chang, Manshik Kim and Kyoungdoug Min. "Detection of misfire and knock in spark ignition engines by wavelet transform of engine block vibration signals". School of Mechanical and Aerospace Engineering, Seoul National University, Seoul, Korea. 20 June 2002

[7] Youngkyo Chung ,Choongsik Bae,Sangmin Choi, Kumjung Yoon. "Application of a Wide Range Oxygen Sensor for the Misfire Detection", International Fuels & Lubricants Meeting & Exposition, May 1999, Dearborn, MI, USA, Session: Diagnostics in SI & Diesel Engines

Francesco Lo Bue, Antonio Di Stefano, Costantino Giaconia
Dipartimento di Ingegneria Elettrica, Elettronica e delle Telecomunicazioni
Università degli Studi di Palermo
Viale delle Scienze, ed. 9 – 90128
Palermo
Italy

Emiliano Pipitone
Dipartimento di Meccanica
Università degli Studi di Palermo
Viale delle Scienze, ed. 9 – 90128
Palermo
Italy

Keywords: misfire detection, spark ignition engine, engine speed analysis, engine speed filter microcontroller

Francesco Lo Bue, Antonio Di Stefano, Costantino Giaconia

"Intelligent" High Pressure Sensor for Automotive Application

A. Müller, A. Fehlinger, T. Pasler, E. Jansen, T. Pasler, M. Totzek, P. Krause, W. Goeser, First Sensor Technology GmbH (FST)

Abstract

Typical markets for high pressure sensors are common rail diesel, gasoline direct injection (GDI), transmission, oil pressure and brake application with their variety from medium up to high pressure ranges like 35, 70, 200, 260, 1800 up to 3000 bar. These markets require a simple, cost-efficient, reliable and robust concept.

Target of the development was a pressure sensor with a minimized number of parts and the usage of state of the art processes for manufacturing. Furthermore, the advanced feature of a stainless steel membrane in harsh environment and a wide application specific flexibility in the mechanical and electrical connector design was also the objective.

The pressure sensor designed by First Sensor Technology GmbH (FST) uses a special patented piezo-resistive monolithic sensor element for measuring the point stress on a steel diaphragm. The sensor works in a temperature range between -40°C and 140°C and has an accuracy of 1..3% depending on the temperature range. With this concept a pressure sensor with minimized outside dimensions has been developed, applicable for engine oil, diesel, transmission oil, gasoline and blend, carbon dioxide and other media in a pressure range between 35 bar and 3000 bar.

The pressure sensor concept has a one part pressure connector, which is made of stainless steel and includes thread and diaphragm. Therefore, no further welding processes are needed and we also have a clear separation from pressurized media. On top of the diaphragm an ASIC including a piezo-resistive sensor element (Weathstone bridge) is soldered on a metal membrane. This ASIC includes all fail safe functionalities like an integrated sensor check of signal conditioning path and others. The sensor element has to be electrically connected by wire bonding to the PCB. The housing with integrated pins is assembled afterwards. The complete pressure sensor is calibrated in an end of line calibration.

In this article the next generation of high pressure sensors for automotive and industrial applications is discussed for high volume markets. One of the most important topics is the fail safe concept and the possibilities and benefits for customers use. Furthermore, the flexibility to accomplish customer needs in combination with short development terms is also discussed. We offer to meet the customers applications like qualifying new materials, modifying outside dimensions, etc. Another topic is the robust design and a description how to withstand harsh environments. Additionally, results of the finite element analysis (FEA) are also shown.

1 Introduction

In a wide range of high pressure applications a clear market growth (e.g. common rail, brake systems, engine oil) combined with new applications can be observed (e.g. CO_2 in new air conditioning systems, measurement of pressure in the engine). In parallel the demands regarding cost reduction, smaller size, high reliability, customized electrical output (digital, analog) and customized mechanical interfaces (plug, pressure connector) are increasing to achieve market advantages in the system.

In the high volume market the mainly used effects for the measurement of high pressure are:

- metal thin film resistors on steel (geometrical effect of strain gauges)
- poly silicon thin film layers on steel (piezo-resistive effect of poly silicon)
- single crystalline resistor elements on steel (piezo-resistive effect of microfused silicon strain gauge)

However, common high pressure sensors suffer from couple of limitations:

- sensor element and ASIC are separated (additional packaging processes necessary)
- special processes are used (e.g. semiconductor processes on steel leading to high production investment and sometimes production costs)
- sometimes very low k-factor (e.g. thin film resistors) requiring high stress in very thin or large membrane geometries and large resistor dimensions
- lots of components are necessary (cost and reliability sensitive)

These key points have to be answered to achieve further miniaturization and a lower price at increased requirements regarding reliability, self diagnostic functionalities and higher temperature ranges.

The FST idea was to integrate all sensor parts as far as possible to reduce the number of parts, connections and achieve a miniaturized space demand. Furthermore only matured technologies should be used to come to high sensor and production reliability. This applies in particular to the used ASIC with integrated fail safe functions.

In this article we describe the feasibility of our concept with Finite Element Analysis (FEA) and measurement results. Furthermore we give you an outlook over the next generation of pressure sensors in harsh environments and our contribution to accomplish customer needs.

2 Product Design

2.1 Design

The ASIC has an integrated piezo-resistive Wheatstone bridge and is grinded down to 100 µm. This ASIC is mounted on a thin glass solder layer on top of the stainless steel membrane. The electrical connection of the ASIC is realized by wire bonding to a carrier board. The signal conditioning includes calibration features for compensation of temperature dependencies and pressure non-linearity as well as for amplification. Programmable clamping levels and fail safe features are included too. For the production of the ASIC and the glass soldering process only mature processes are used.

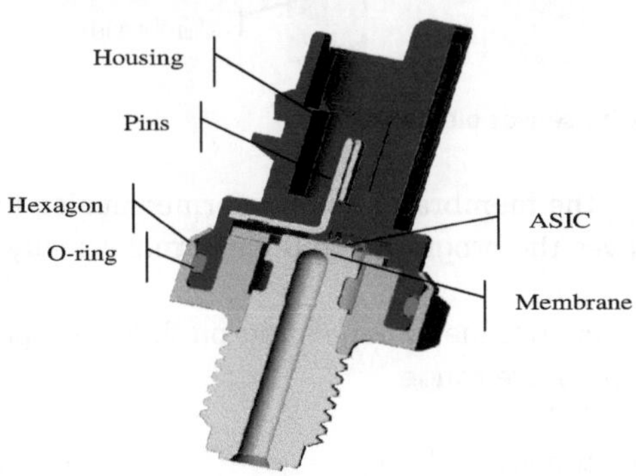

Fig. 1. Cross section of FST high pressure sensor

For the soldering process a temperature within allowed annealing temperatures for ASICs below 450 °C is used. A main topic at FST was the development of the glass soldering process. High process stability has been reached avoiding any kind of offset shift. As membrane material a standard stainless steel is used. The high pressure sensor (Fig. 1) consists of a housing made of PBT 30 GF with integrated contact pins. The housing is assembled to the Hexagon by joining by flanging. To fulfill a degress protection according to IP69K an O-ring is assembled additionally.

2.2 Functional Description of Sensor Element

The pressure sensitive part of the sensor consists of a steel membrane of a certain thickness and an ASIC mounted on top. The ASIC includes a Wheatstone bridge to detect stress changes. The layout of the Wheatstone bridge (Fig. 2) is zoomed out. There you can see the structure of the piezo-resistive elements. Normally the Wheatstone bridge is mounted in the elongation of the bore hole edge to achieve maximum sensitivity. The structure of the bridge resistors is longitudinal and transversal. Because of the small ASIC-dimensions a redundant ASIC on the same membrane is possible.

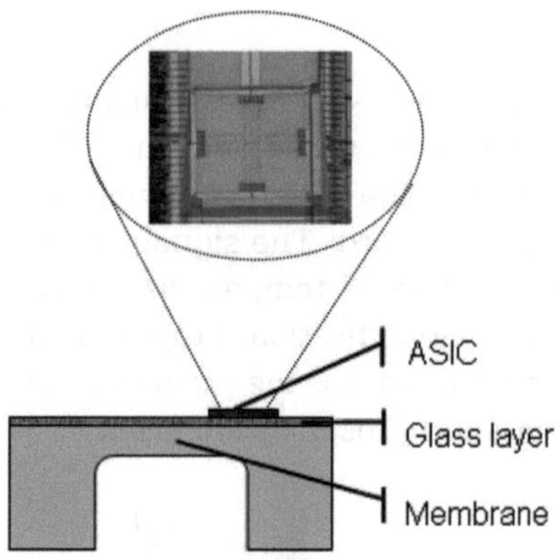

Fig. 2. Structure sensor element

In case of $p_0 = p_{ambient}$ the membrane is not deformed analog to Fig. 3. The output of the sensor gives the programmed 0 bar signal (usually 0.5 V).

The sensitivity of a membrane is depending on following parameters and is individual for each pressure range.

- diameter bore hole
- membrane thickness
- radius of bore hole end

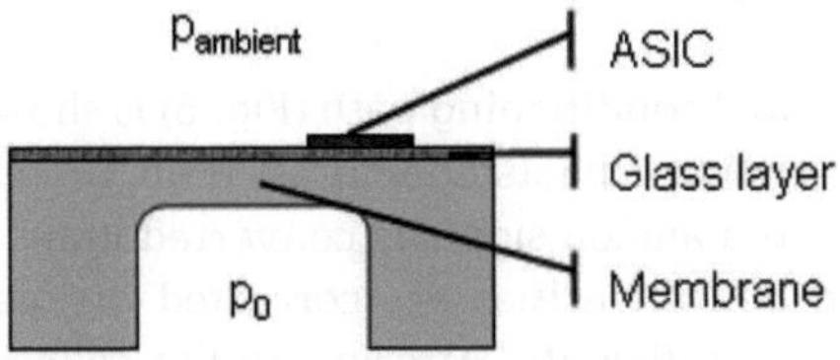

Fig. 3. Membrane at $p_0 = p_{ambient}$

In case of $p > p_{ambient}$ the principle deformation of the membrane is according to Fig. 4. Under pressure the membrane is deformed. Due to deformation mechanical stress is transferred into the piezo-resistive Wheatstone bridge. The special design of the T-bridge allows the concentration almost in one point. The bridge is located at the membrane edge. Therefore the same ASIC can be used for different pressure ranges independent of the hole diameter. Similar to standard piezo-resistive silicon pressure sensors the output voltage is direct proportional to the pressure value. Thereby only one ASIC is necessary over all pressure ranges, the required adaption to pressure ranges are done by modification of the membrane geometry.

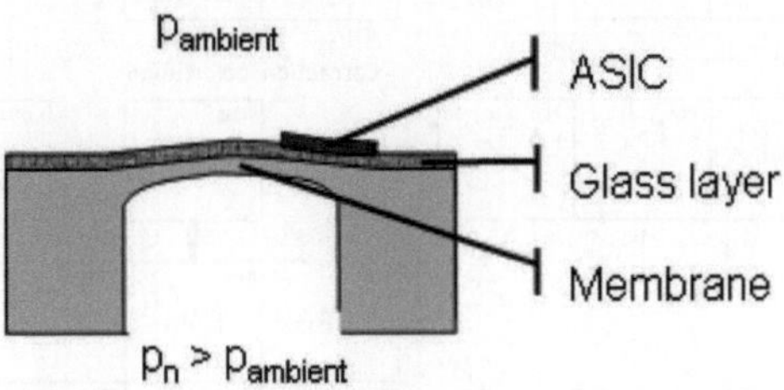

Fig. 4. Membrane $p > p_{ambient}$

Because of the standard temperature dependencies (offset, sensitivity) and non-linarities standard ASICs for piezo-resistive sensor elements can be used. This allows us also to realize a smart fast calibration. The ASIC also compensates assembly dependent differences in membrane geometry, ASIC- and packaging parameters.

2.3 Signal Conditioning

A schematic of the signal conditioning path (Fig. 5) is shown below. The output signal of the piezo-resistive Wheatstone bridge is amplified with a programmable gain factor. Then the analog signal is converted into a digital signal where zero point shift and non linearities are corrected by specific and individual calculated calibration coefficients. Afterwards the calibrated digital signal is reconverted into an analog output voltage signal. Digital output signals are also possible depending on customer requirements. A typical analog output signal is also shown below (Fig. 6).

FST uses a customized ZMD-ASIC for piezo-resistive sensors with the integrated FST T-bridge. Detailed experiments show that the stress influence on the ASIC caused by packaging and membrane deflection can be neglected.

Furthermore temperature measurement with a diode is also possible.

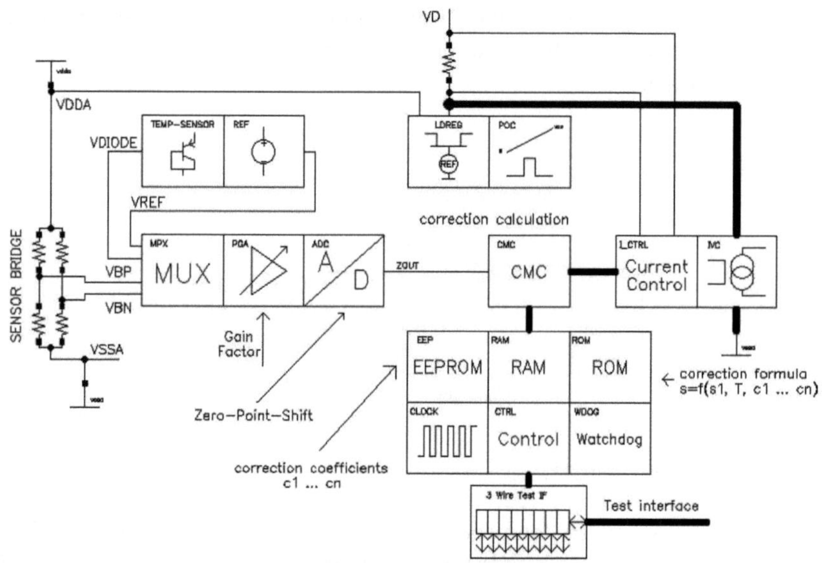

Fig. 5. Schematic of signal conditioning path

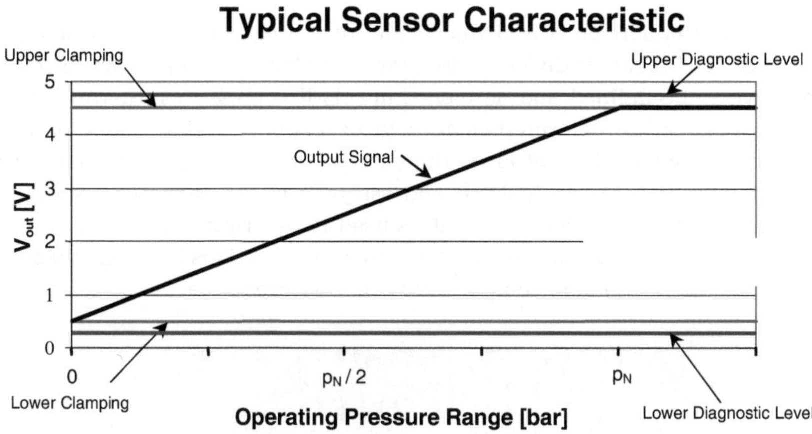

Fig. 6. Typical output characteristic

2.4 Finite Element Analyses (FEA)

An important task of the sensor design is to find optimized geometric parameters for the different components. An optimized shape of several parts as well as the entire device is crucial to achieve a high signal to noise ratio and long term stability as well as to get a robust design allowing a cost effective production.

We performed finite element analysis to investigate the influence of different materials, part dimensions and sizes. Thus a time consuming making and testing of prototypes could be avoided.

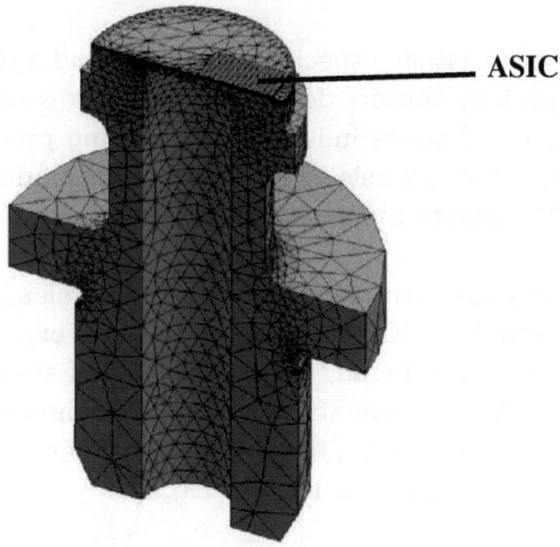

Fig. 7. Meshed model of the sensor carrier

Basis for all simulations is a virtual model of the sensor, including material properties. The model is divided into a mesh of elements (Fig. 7). Basic conditions have to be defined and ambient impacts like pressures, thermal treatment or mechanical deformation have to be applied to the model. Solving equations following physical laws, the remaining field of physical data, e.g. stress or temperature is calculated and assigned to the elements. Especially the stress on the top side of the ASIC is used to determine the output of the sensor bridge and therefore the sensor signal. (Fig. 8). This has been taken to verify simulation results in comparison with measured data.

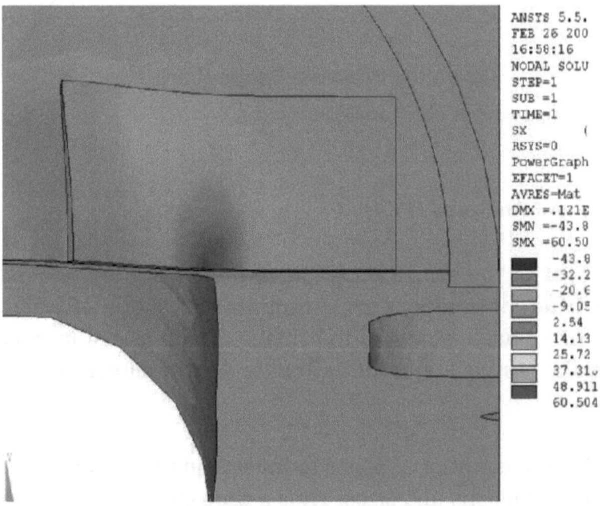

Fig. 8. Stress distribution in the ASIC under an applied pressure. Ideally, the bridge position fits with the position of maximum stress.

Beside the usual optimization a special task was to find a stress-reducing form of the carrier. Especially sensors designed for low pressures tend to be very sensitive to any kind of stress induced by mounting processes. In case of a 35 bar sensor with no stress releasing features the simulation gives a signal shift of about 100% created by unwanted stress (Fig. 9)

In order to eliminate any disturbing signal, a stress releasing shape has been integrated. Using standard DOE techniques (design of experiments), a geometric parameter set has been found, where any kind of stress generated at the fixed part of the carrier does not affect the ASIC, no matter if it is of mechanical or thermal origin. For a stress optimized 35 bar sensor the remaining disturbing signal is in the range of 0.1% (see Fig. 9).

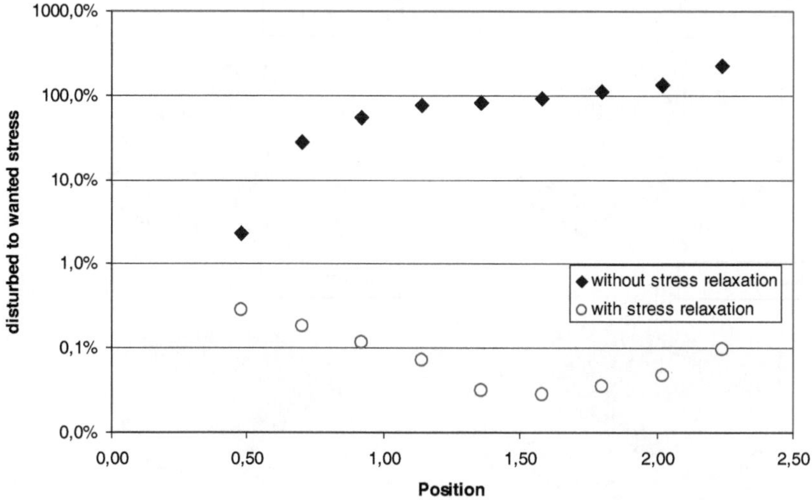

Fig. 9. Comparison of optimized and not optimized structure

3 Measurement results

3.1 Performance

Following figures show typical results (typically error in %FSS) of the p-T-performance measurement for several nominal pressure ranges. The p-T-performance starts at 20°C, rising in 20°C steps up to 140°C and decreasing with the same stepwidth down to -35°C and back to 20°C. At each temperature-step the pressures are applied from 0 bar to nominal pressure and back to 0 bar with a stepwidth of 10% $p_{nominal}$.

An air shock test was performed with 10 samples and 1000 cycles – all samples passed this test.

A shortened pressure cycling test about 2 million cycles with enhanced maximal pressure of 2300 bar was performed with 10 samples of the high pressure Sensor -1800 bar-series – all samples passed.

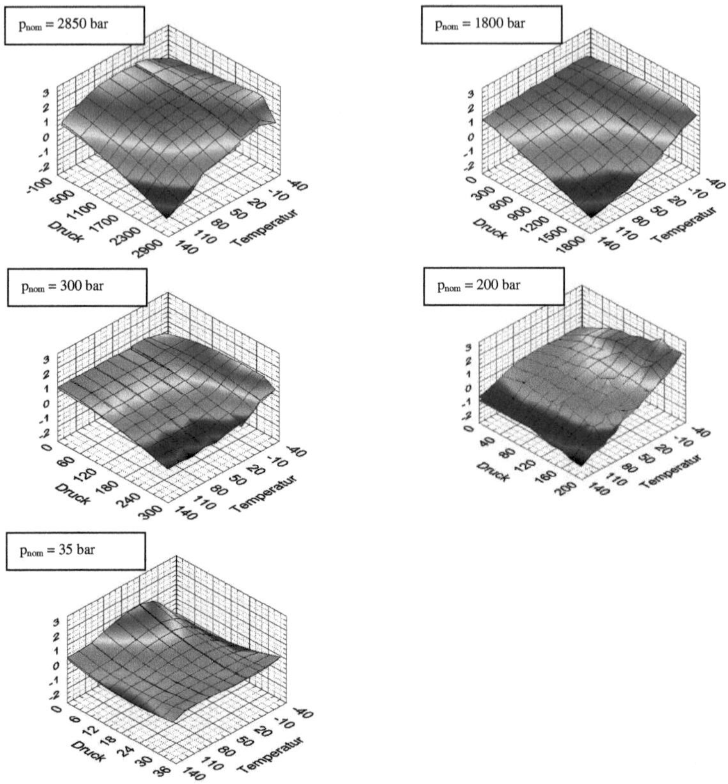

Fig. 10. overview of pressure ranges and measurement results

3.2 Stability of Output Signal

Following figures show the stability of output signal about a couple of p-T-performance measurements. Between the measurements the samples were stored under different storage conditions (e.g. air shock and stored at -50°C).

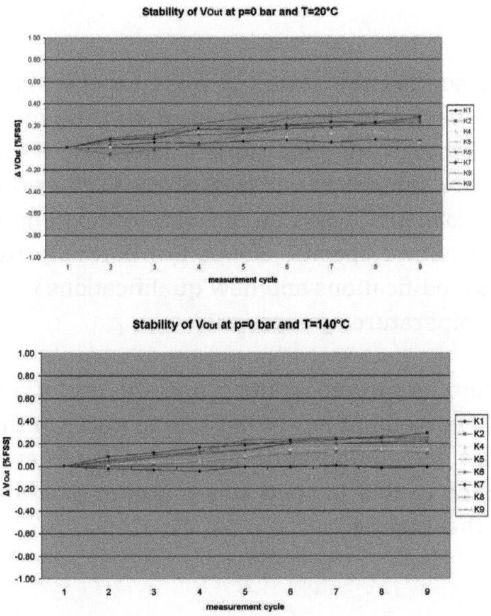

Fig. 11. Typical stability at $p = 0$ bar @ 20°C and 140°C

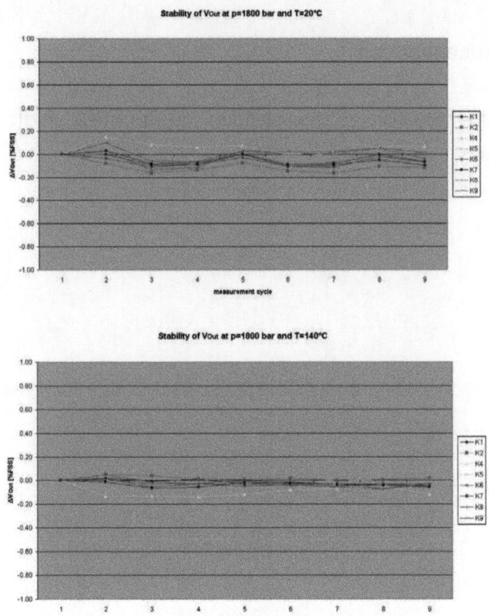

Fig. 12. Typical stability at $p = 1800$ bar @ 20°C and 140°C

5 Conclusions

In this article we described the status of work for the high pressure sensor. The objectives have been completely achieved and the measurement results and FEM simulation show very good results. Therefore we are able to offer a wide range of pressure sensors to our customers. Furthermore we are able to develop customized concepts based on our core competencies in short terms. The variance and customer specific claims are normally the fitting and the housing. With short modifications and new qualifications we are able to create solutions for high temperature applications.

Because of a complete integrated solution and the use of a customized ZMD-ASIC, we are able to implement lots of fail safe features in the ASIC. The measurement of pressure and temperature allows us also to create an intelligent pressure sensor which is able to send status information about pressure and temperature to another system.

Andreas Müller
First Sensor Technology GmbH
Forschung und Entwicklung
Carl-Scheele-Str. 16
12489 Berlin
Germany
andreas.mueller@first-sensor.com

Keywords: pressure sensor, piezo-resistive, automotive, high temperature, high pressure

Networked Vehicle

Embedded Security Solutions for Automotive Applications

A. Bogdanov, D. Carluccio, A. Weimerskirch, T. Wollinger, escrypt GmbH

Abstract

In this paper we present a number of architectural security solutions based upon concrete hardware components such as customized security controllers, trusted platform modules (TPMs), "security boxes", FPGAs and ASICs. We analyze benefits and disadvantages of each solution proposed in terms of physical and cryptographic security, costs, needed and achievable performance. We also discuss the consequences of the solutions with respect to several wide-spread security applications including immobilizer systems, component identification, software flashing, etc.

1 Overview

During the last years the significance of information security in embedded systems has drastically increased. This is due to the fact that embedded systems are getting more and more pervasive. In the context of their growing networking degree embedded security becomes particularly important.

One of the most rapidly developing areas where information security plays a considerable role is the automotive industry. A modern automobile has up to 80 embedded micro controllers (control units - CUs) on board which control most processes in the car (e.g. engine control, steering and braking systems, navigation systems, traffic control, etc.), and build a heterogeneous, multi-rank communication network with broadband interfaces to open computational environments (see Fig. 1). The cost for electronics and software is approaching 30% of all manufacturing costs.

Security solutions based on software approaches have been gaining attention in the automotive industry throughout the last years. These are vulnerable to attacks based on malicious software, though. Recently a discussion about a secure hardware component as security anchor has been started. However, it is unclear yet how secure computing modules are able to add security to a vehicle, and which security solution should be preferred. In this white paper we give some initial thoughts about these aspects in order to start a discussion about future approaches.

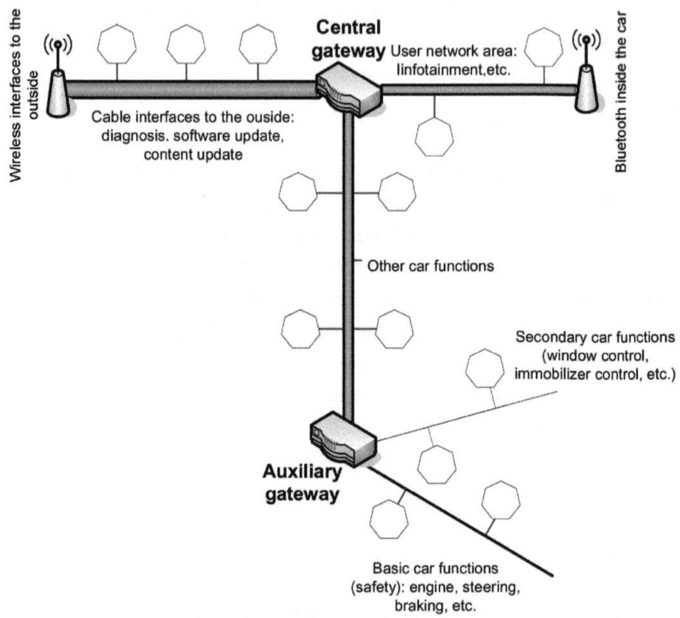

Fig. 1. General in-vehicle network topology

2 Secure Hardware Computing Base

Providing security in embedded systems is a challenging task requiring an interdisciplinary approach encompassing cryptographic methods, technical security solutions and organizational measures which should be taken to provide an adequate security level. Traditionally, the majority of high-security solutions in the embedded area are based on one of the following secure computing platforms.

2.1 Security Controllers

Security controllers are special microprocessors protected against active (tampering and other invasive attacks) and passive (timing attacks, simple differential power analysis, internal collision attacks, EM analysis, template attacks and many others) physical attacks. They offer a number of pre-implemented cryptographic services such as DES/3DES, AES, hash functions, finite field and residue ring arithmetic, RSA, secure generation of random numbers (needed for key management, some signature schemes, security protocols, etc.) and others. These cryptographic functionalities are usually implemented as co-processors. Such controllers are also able to store data in secure area. Hence, usually the data can be written once, but can afterwards only be read out or only used by the security controller for cryptographic operations. Most of these controllers are smart card derivations delivered in traditional micro controller packages (e.g. DIL, TSSOP, DSO, etc.). Such security controllers possess 8-, 16- or 32-bit central processing units with clock frequencies between 8 and 66 MHz, 2-16 kB RAM, 16-256 kB ROM, and up to 400 kB EEPROM. There are, however, security controllers with larger (up to several MB) EEPROM and ROM.

The advantages of such controllers are as follows:

- These are special-purpose high-security solutions whose hardware and firmware architectures are well-understood and whose security has been as a rule thoroughly evaluated and certified by state certification bodies within formalized certification procedures (e.g. by Common Criteria (CC) [2] Protection Profiles (PP) [3, 4]).
- They are already available on the market and can be relatively quickly produced after developing the corresponding embedded software application.
- relatively low manufacturing costs (their price is in the range of one to several Euros for high volumes).

At the same time there are several technical drawbacks impeding their potential wide adoption in automotive applications:

- relatively low computational performance in view of the need of real-time reaction in safety-relevant applications.
- relatively low data transmitting capabilities which may prevent standard security controllers from controlling broad-band in-car communication and external interfaces on-line.
- relatively narrow range of operational conditions. The allowed temperature range is as a rule from -20°C to +85°C which is to be compared to the required values for automotive applications from -40°C to +105°C. For solving these problems some modifications in the hardware core through semiconductor manufacturers may be necessary.

2.2 Trusted Platform Modules

A special type of smart-card based security controllers are so called Trusted Platform Modules (TPMs) whose security functionality is defined by the TCG (Trusted Computing Group) standardization body in [5]. Originally TPMs were developed to provide platform integrity (software and hardware integrity) of PCs with emphasis on software attacks. The security services provided by TPMs and their interfaces are standardized.

The basic functionalities of a TPM are the following:

- secure storage of cryptographic keys and hash values (representing platform configuration)
- on-chip key pair generation using an unpredictable hardware-assisted random number generator
- computation of hash-values
- signature generation and verification
- monotonic counter

TPMs have clear advantages:

- These are ready products wide-spread on the information security market.
- As a rule they are CC certified by their manufacturers.
- TPMs use clearly specified, well-understood and transparent security mechanisms.
- They are relatively cheap (in the range of one Euro).

But their drawbacks are much more considerable:

- TPMs offer a limited set of cryptographic services which is often not sufficient to build a multi-functional information security system apparently required in the automotive area.
- The available functionality is fixed by the TCG and chip manufacturers and cannot be extended.
- To securely use their functionality it is almost always necessary to have a trusted computation environment (or some elements of it) outside the TPM.
- Moreover, all limitations of security controllers mentioned in Section 2.1 persist.

2.3 Security Boxes

Security boxes are high-performance computer devices additionally protected against physical attacks (tamper and side-channel resistance). They can use standard high-end microprocessors as well as FPGAs/ASICs and possess a performance level comparable to modern PCs and higher. Security boxes are deeply customized products and available at the moment mainly in military and state security applications (e.g. communication devices installed in embassies, army headquarters, fighter aircrafts, and anti-aircraft defence emplacements). The concept of security boxes found limited application in some industries as well (e.g. measurement in the trunk distribution of electricity and energy carriers).

Secure boxes have a number of benefits valuable for the automotive industry. The most important ones are:

- high computation (almost arbitrary hardware base can be taken) and communication (among others because of the possibility of hardware-assisted realization of communication protocols) performance
- high performance of cryptographic algorithms, since they can be realized in hardware which is as a rule much more efficient
- high storage capacity

All these properties enable security boxes to perform centralized online control of automotive security features.

The only relative disadvantage of this approach is that there is neither a solid development base nor ready solutions available. That is, security boxes must be developed from the outset, which means high development costs. The same applies to certification procedures (e.g. there exists no Common Criteria protection profile for automotive security boxes). Moreover, security boxes will be relatively costly in manufacturing.

3 Security Objectives

The security issues in vehicles usually comprise the following topics:

- electronic immobilizer
- remote entry system
- secure software download (ensure software integrity)
- secure access (for diagnosis purposes and further external channels)
- digital rights management (infotainment)
- mileage counter manipulation
- component identification (ensure hardware integrity, e.g. detects theft and forgery)

Some examples for future scenarios including security are:

- car-to-car communication

- car-to-infrastructure communication
- data event record
- X-by-wire
- speed control
- toll collection
- secure internal communication

Concerning data, software and hardware manipulation there are two security levels to achieve:

1. **Avoid manipulation:** Every state of the system is permanently controlled by the security mechanism in real time which guarantees that the system is in a secure state at any time moment.
2. **Detect manipulation:** If the system has been once in an insecure state, the security mechanism will detect that within a finite number of steps after this event. Practically, we are interested in a security mechanism detecting manipulation directly at the time of manipulation or immediately after this.

While it is desirable to secure crucial vehicle components such as the engine this seems to be out of scope today. Hence, we focus on securing the CUs connected to an internal bus. The overall security goals can easily be summarized to avoid and detect manipulation, respectively, in order to achieve software and hardware integrity. Concerning data confidentiality, one can speak of reliably providing confidentiality or failing to provide it. Moreover, similar classes of avoiding and detecting information leakage can be introduced.

4 Security Architectures

There are three basic security architectures:

- a centralized approach where all security is handled in a centralized way (client-server architecture),
- a distributed approach (peer-to-peer architecture),
- a semi-centralized approach which is a combination of the centralized and distributed architectures.

However, each concrete approach must be mapped to hardware solutions (standard controllers, TPMs, customized security controllers, security boxes).

4.1 Centralized Architecture

In the centralized approach there is a single security module that is responsible for the overall security. This module has to withstand State-of-the-Art attacks and has to perform the necessary security operation in an acceptable speed. However, the required performance depends on software architecture and is application specific.

Each CU shares a relationship with the security module (e.g. an individual secret key). The security system can in this case realize a server-based authentication and key-establishment protocol, the security module (SM) being the protocol trusty. If a CU_i wants to build a secure channel to CU_j, it asks SM to participate in establishing a session key between these two CUs. After establishing a key, the both parties can communicate with each other not involving SM. The CUs in this case have neither to maintain all cryptographic keys of their communication partners nor to store any master keys in their potentially insecure computational environment. This considerably reduces the computational power required of SM to maintain in-car communication sessions. There is, however, still the necessity to be able to distribute all necessary keys fast enough to support real-time (e.g. at the start-up time) and safety critical (e.g. braking) applications. The SM has to store only a single car master key. It can derive keys needed to communicate with CUs from this single key and the key storage requirements can be significantly reduced. However, the communication to the outside should be performed by SM directly without delegating this to a less protected CU. In this way all out-car communication sessions can be recorded. Moreover, the SM can also incorporate several security relevant CUs such as electronic immobilizer, security relevant parts of the engine control unit and mileage counter.

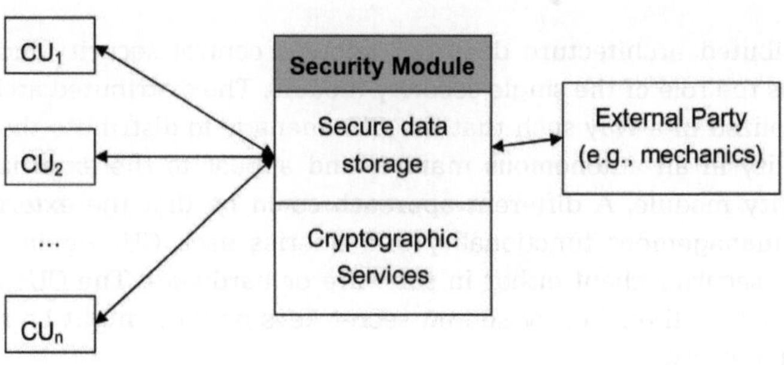

Fig. 2. Centralized architecture

The security module might be a dedicated CU or it might be part of a CU that already exists (such as the head unit, engine control unit or central gateway). There might be a communication channel between the security module and an external entity, e.g. the OEM or a mechanics via a diagnosis channel. This architecture is depicted in Fig. 2.

Note that the security module is the central point of failure and attack. Hence, compromising the security module must be at least as hard as compromising all (or a considerable part of) CUs. If S denotes the security level, then it must hold: $S(SM) \geq S(CU_1) + S(CU_2) + \ldots + S(CU_n)$, where n equals to the number of CUs in the car.

In the centralized architecture the security module may be realized on the following bases:

- **Standard non-security processor:** This solution can be acceptable in rather unlikely cases where physical attacks are no concern (e.g. if malicious physical access to the car is securely protected by organizational measures). Even in this case the processor has to be powerful enough to maintain key establishment and fast communication to the outside.
- **Security controller:** This can be a solution of choice, provided the security controller is made robust to the automotive environment and supplied with fast and secure cryptographic co-processor to maintain high-speed out-car communication.
- **Security box:** This hardware solution seems to be the most suitable one for the purpose of central security unit. It can undoubtedly provide both real-time and out-car protected data transmission and control.

4.2 Distributed Architecture

The distributed architecture does not know a central security module but distributes the role of the single security module. The distributed architecture can be realized in a way such that the CUs manage to distribute the security functionality in an autonomous manner and appear to the external unit as one security module. A different approach could be that the external party provides management functionality and queries each CU. Again, each CU requires a security client either in software or hardware. The CUs might be related to each other, e.g. by shared secret keys or they might be related to the external party.

In the easy case the external party shares a secret key with each CU. Security is then based on this relationship and the possibility of the external party to communicate with each CU. This would mean high storage complexity for each individual CU and the limitation of the security of the whole system to the security of the weakest CU (or a set of the weakest CUs needed), since each CU must then contain all keys it needs. Otherwise the key logistics would be to complex. In the second case all distributed CUs provide an abstract single security module to the external party. In such case the key management comes at far lower overhead.

However, such an approach needs to be more researched since several aspects of its implementation are unclear today. The distributed architecture is depicted in Fig. 3.

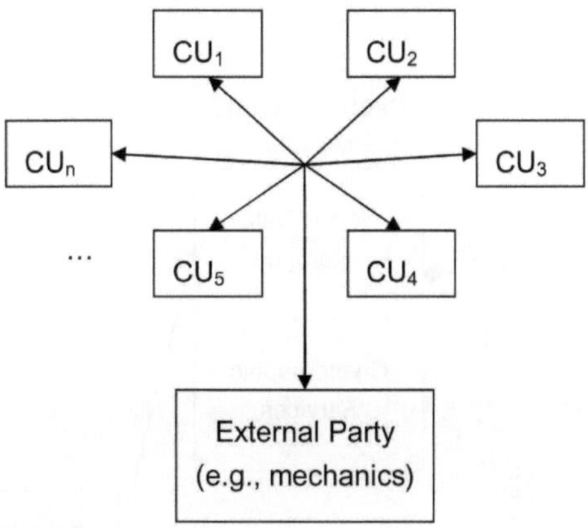

Fig. 3. Distributed architecture

As to possible concrete hardware realizations, the following solutions seem to be most appropriate:

- **Standard non-security processors:** In this case system designers have to come up with a distributed key establishment system possessing suitable performance for real-time and safety critical applications.
- **Several security controllers:** For the most security demanding applications several security controllers are used, the other CUs being based on standard non-security processors.
- **Several TPMs:** Some applications requiring only limited security functionality can be realized using TPMs as well. A TPM is, however, dependent on its owner that calls its functions from the TPM and needs to be implemented in a secure environment.

4.3 Semi-Centralized Architecture

The semi-centralized architecture incorporates security features of both centralized and distributed approaches: There is a central security module supporting security functionality of most CUs. At the same time, some of the CUs (probably, the most security demanding ones) are realized as separate security modules with their own communication channels to the outside world (Fig. 4). These separate security modules might have connections to the central security module and/or to the CUs over their own key establishment system (this would increase the key storage complexity in each CU only linearly). Potential candidates for the separate security modules are engine control unit, electronic immobilizer, and mileage counter.

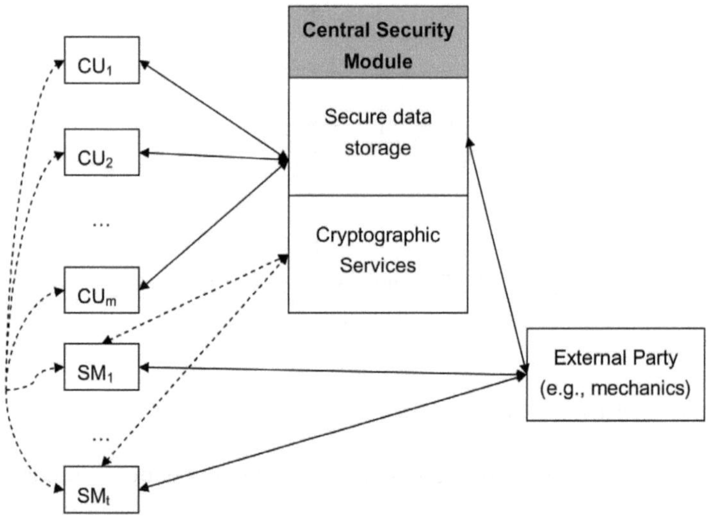

Fig. 4. Semi-centralized architecture

The most reasonable hardware solutions for the semi-centralized architecture include:

- **Standard non-security processors (CUs), security controllers (SMs) and a security box (CSM):** Applications that do not require a high security level (e.g. wind shield wiping controller, sitting controller, etc.) can be implemented on standard microprocessors. The corresponding software should however support basic security functionality. Some security relevant applications can be implemented in security controllers which can delegate computationally intensive tasks to the central security module based on a security box.

- **Standard non-security processors (CUs), security controllers (SMs/ CSM):** The communication between individual CUs (standard non-security processors) and SMs (security controllers) is controlled by a central security module which is based on a security controller. The CUs can also use some security functionality of the central security controller.

5 Security Analysis

In the following we will analyze above architectures in more detail. As already mentioned the autonomous distributed approach requires further research such that we believe only the centralized architecture and the simple distributed approach to be deployed in the near future.

It seems clear that any economically rational security strategy is not able to **prevent** an attack based on hardware manipulation. For instance, it is always possible to replace crucial CUs by malicious components that provide the exact same functionality without the security functions. A proper example is a CU that does not verify the electronic immobilizer. The major security objectives one can achieve are as follows:

1. **Avoid** manipulation due to attacks based on malicious software.
2. **Detect** attacks based on manipulation of hardware.

The first aspect can be ensured by defining a proper interface from the outside world to the vehicle and by protecting this interface. For example, standard access control and authentication of any data imported to or exported from the vehicle will allow such an approach. Furthermore, for critical components such as the engine control unit the standard interfaces can be protected by cryptographic means. For instance, the flash boot loader should be appropriately protected. However, protecting such devices of direct hardware manipulation, e.g. by direct physical contact, is only possible to a certain point. It is possible to protect the flash memory of direct contacting and manipulation by using a dedicated security chip. However, it seems almost impossible to prevent an attack where an attacker just introduces a further controller that partly takes over the role of the old one. For preventing attacks based on software there is no need for a dedicated hardware security module that incorporates features of a security chip but the functionality might be simply implemented in software on a standard controller. This goal can be achieved with all architectures in question (centralized, semi-centralized, and distributed). The main issue to consider here are the interfaces to the external world and how they are protected.

The second major goal of detecting a hardware manipulation can be achieved at a high probability but must be thoroughly planned. The main idea is as follows:

- **Centralized and semi-centralized security architectures:** The central security module queries the CUs at determined events and records all responses. It applies plausibility checks to the results and is able to detect manipulations. The security module might directly react to a security breach or notify an external party (e.g. once a diagnosis check is performed).
- **Distributed security architecture:** Each CU records its events. An external entity (say, an OEM trusted mechanics) is able to read out all these records, e.g. at the time of a diagnosis check. The external party than applies plausibility checks and is able to detect irregular behavior at a high probability.

The plausibility checks depend on the characteristics of the CU. For instance, the following plausibility checks might be applied for different CUs:

- **Mileage counter:** The security module can store a copy of the mileage counter or it can store the actual mileage counter. Thus, the sensors would send the signal to the security module which then increases the mileage counter appropriately. If the security module has secure storage the counter cannot be altered once it is stored in the module. An attacker would have to alter the communication to the security module. However, the module could check whether the signal is disrupted (e.g. by checking whether the car is turned on for a long time without any sensed driving) or if it is altered (e.g. by checking whether the car is always going very slow). A trusted third party such as the OEM could then read out the records and detect the manipulation.
- **Electronic immobilizer:** The role of the electronic immobilizer is to ensure that a vehicle can only be turned on if the appropriate vehicle key was used and if the appropriate CUs are present in the car. Usually, if the check is successful the engine control unit is activated and the engine can be started. An easy attack is to just replace the engine control unit and, if necessary, further crucial CUs. The security module could record the responses of the security check when querying all involved CUs. If the CU was replaced, there would be no proper response of the key and the engine control unit, but the security module would detect that the vehicle is still running. A trusted third party could then detect the attack.
- **Component identification:** Component identification [1] ensures hardware integrity of the vehicle. It is in general very similar to the mechanisms of the electronic immobilizer, which verifies hardware integrity of the key and dedicated CUs. Hence, the same mechanisms can be used here.

A common attack will be to remove or replace the security module, or to add a second module that reacts to the other CUs in the same way. However, with a plausibility check and the knowledge of the secret keys of the security module, a trusted third party is able to detect such a manipulation.

Note that detection of hardware manipulation can only be detected up to a certain degree. An adversary that is aware of how the plausibility check is performed and that has sufficient monetary resources is always able to compromise the system. However, we believe that it is possible to design plausibility checks in such a way that an attack's effort is higher than the gain such that no rational adversary will implement such an attack.

We started this article with a description of TPMs. But how well suited are TPMs for the application in vehicles?

The pros and cons of a TPM to a custom specific solution (e.g. an ASIC or an FPGA solution) are summarized in the following table:

	TPM	*Custom specific*	
		ASIC	*FPGA*
Standardized	*Yes*	*No*	*No*
Flexibility	*No*	*Yes*	*Yes*
Prize	*Medium*	*Low for high volumes*	*High*
Security level	*High*	*Adaptable to required level*	*Medium - high*

Tab. 1. Pros an cons of a TPM and a custom specific solution.

Finally, one can say that there is no winner. TPMs have the clear advantage of being a standardized security module with high volumes. On the other side, they come with an interface that was designed for the PC world. There are approaches to adapt TPMs to the mobile and vehicle world though. A custom specific solution can be exactly suited to the requirements of vehicle OEMs regarding functionality and required security level. However, an appealing prize seems only possible if several OEMs agree on a standard security module such that high volumes can be reached. An FPGA solution seems only reasonable for small volumes due to its prize.

6 Outlook

In this report we analyzed using a security module in vehicles. Clearly, several issues need to be resolved first. For instance, the security module needs to hold the stress and temperature requirements in vehicles. However, the chances for a security module are appealing:

- a single security module might safe code size and hence even cost, because one needs less memory, processor power as well as developing cost
- Software attacks can be prevented and hardware attacks can be detected, hence software as well as hardware integrity can be proven by a trusted third party.
- Communication to the vehicle can be protected by the security module (e.g. as part of a central communication gateway for diagnosis, wireless communication as well as further channels) such that introduction of malicious communication is detected and has no affect.
- Business models requiring a secure digital rights management (DRM) can be introduced (e.g. aftermarket feature activation and custom specific use of digital contents).

However, the main advantage of a hardware security module compared to a software solution is that hardware attacks can be detected and that a dedicated security module is available that is able to perform cryptographic operations much faster and with little resources even compared to powerful 32-bit standard controllers such that in the end cost might be reduced.

The design of a practically secure and highly functional automotive computer system resistant to all software attacks concerning confidentiality and manipulation as well as some powerful physical attacks is possible. The main candidates for security architectures and secure hardware platforms are the centralized architecture (based upon a security box or a high-end security controller with a powerful cryptographic co-processor) and semi-centralized architecture (on the basis of several high-end security controllers possibly accompanied by either a powerful cryptographic processor or a central security box).

References

[1] André Weimerskirch, Christof Paar, and Marko Wolf, "Cryptographic Component Identification: Enabler for Secure Inter-vehicular Networks", 62nd IEEE Vehicular Technology Conference, September 25-28, 2005, Dallas, TX, USA.

[2] Common Criteria for Information Technology Security Evaluation. Parts 1, 2 and 3. Version 2.2. January 2004

[3] Smartcard IC Platform Protection Profile BSI-PP-0002, Version 1.0, July 2001

[4] Protection Profile Smartcard Integrated Circuit PP/9806, Version 2.0, September 1998

[5] TCG TPM Specification, Version 1.2, Revision 94, 29 March 2006

Andrey Bogdanov, Dario Carluccio, André Weimerskirch, Thomas Wollinger
escrypt GmbH- Embedded Security
Lise-Meitner-Allee 4, D-44801 Bochum
Germany
abogdanov@escrypt.com
dcarluccio@escrypt.com
aweimerskirch@escrypt.com
twollinger@escrypt.com

Keywords: automotive systems, embedded systems, embedded security, hardware architectures, safety, automotive security.

Automotive 1 Gbit/s Link opens New Century in Car HMI and Driver Assistance Systems

A. Krepil, Inova Semiconductors GmbH

Abstract

Car Telematics systems are becoming the mobile communications platform of choice for cars and trucks. The availability of brilliant, high-resolution TFT-LCD displays, up to 1280x480 pixels and real-time camera systems up to VGA at 60 fps are providing increased capabilities for navigation, office-like tasks and communication, and automated driving options i.e. lane departure warning, obstacle detection, sign recognition, blind spot detection or parking assistance. APIX, Inova Semiconductors' new automotive Pixel link, is uniquely designed to overcome the bandwidth-distance limitations of the physical layers of today's automotive multimedia links. The new 1 Gbit/s link just needs two copper wires (or four wires for bi-directional functionality) to connect displays or cameras with their respective processing units up to 15 m or more. Thus the new link has the potential to enable new driver assistance systems to increase the throughput and to reduce connectivity cost, cable weight and cable diameter.

1 APIX - Advanced Pixel Link

High-resolution TFT LC display links are very bandwidth-hungry and therefore urgently require Gbit/s to/from their control- and multimedia sources. Due to lack of dedicated high-bandwidth physical layers, automotive system designers were already considering placing more fiber cables inside cars. Because of fiber links limited reliability and mechanical robustness, car makers have placed more emphasis on copper-cable based, bit serial communication techniques to save cost and installation headaches. APIX is a long-distance high-speed bit serial link, DC-balanced, low latency and low EMI interconnected, directly, to dispatched displays or cameras connected to central video/imaging processors using only two wires. The integrated bi-directional sideband channel offers an 18 Mbit/s capability that can be used, as an example, for I^2C-compatible operations for controlling CMOS camera sensors or display settings. Even CAN-bus can be transferred almost transparently.

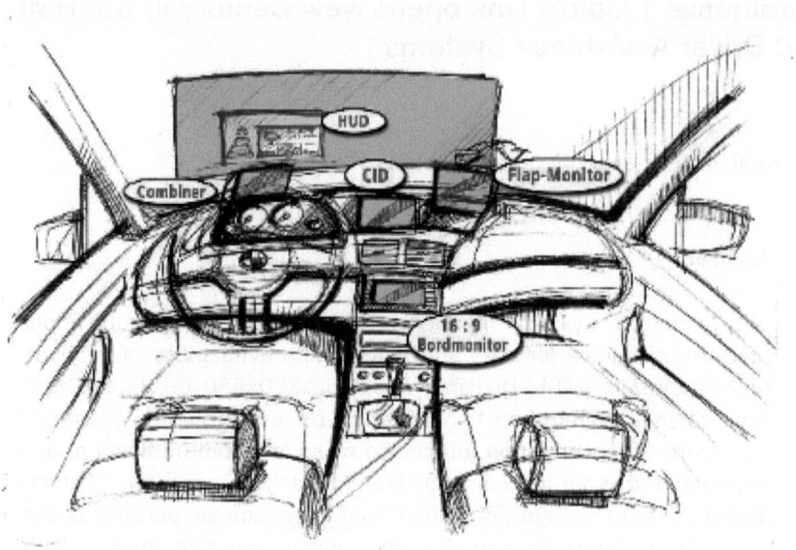

Fig. 1. In-car display applications (source: BMW group)

Selectable 0.5 or 1 Gbit/s operation, adjustable driver characteristics of the CML-IOs (Current Mode Logic) and integrated spread-spectrum-clocking allows for the lowest EMI at maximum transmission distances. This significantly reduces the link's system costs while increasing the system functionality compared to simple Serializer/De-Serializer chip sets. The APIX link chips offer an integrated digital RGB video interface, directly, making it easy to adapt to all small to mid-size LCD-TFT displays with a TTL-interface. Due to a 850 Mbit/s net video data rate, the link serves all SVGA displays (800x600 pixels, 24 colour bits), SWVGA (1280x480 pixels, 18 color bits) up to 25 m distance through CAT-cables.

To achieve maximum link bandwidth at different interface widths (10, 12, 18, 24 bits + HSYNC, VSYNC, PxCLK, DE), the link is configurable during power-up. Additionally it is possible to choose between two physical layer return paths: the embedded (common mode jump signalling on the video down stream link) or the separate return channel (CML). All bi-directional sideband channels offer real-time operation and therefore are totally independent from the pixel clock fed into the video interface. The individual drive current control and Pre-Emphasis adjustment, dynamic line coding and spread-spectrum clocking enable best-in-class EMC performance, especially in complex cable and connector assemblies with non matching constant impedances.

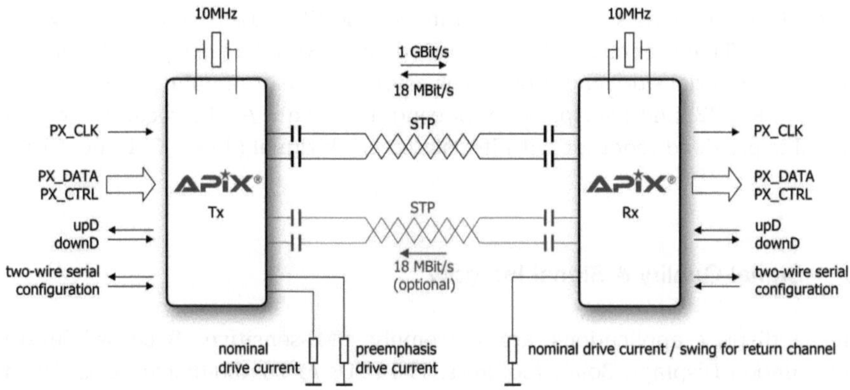

Fig. 2. APIX link overview

2 Link Bandwidth

The APIX link targets automotive LC display and camera applications. To maximize transmission distance and to minimize EMI the link can be set into "Full Bandwidth Mode" (1 Gbit/s) or "Half Bandwidth Mode" (500 Mbit/s). In Half Bandwidth Mode the transmission frequency is just 250 MHz and therefore distances of 25 m+ can be achieved by using simple STP CAT cables.

Channels	*Full Bandwidth Mode (@ 10 MHz Ref. Clock)*	*Half Bandwidth Mode (@ 10 MHz Ref. Clock)*
Downstream	*Max pixels_clk*	*max pixels_clk*
10 bit pixels_data	*62.0 MHz*	*31.0 MHz*
12 bit pixels_data	*61.0 MHz*	*30.5 MHz*
18 bit pixels_data	*42.0 MHz*	*21.0 MHz*
24 bit pixels_data	*32.0 MHz*	*16.0 MHz*
Sideband Channel	*2 x 1 bit @ 12 Mbit/s max.*	*2 x 1 bit @ 6 Mbit/s max.*
Upstream		
Sideband Channel	*2 x 1 bit @ 9 Mbit/s max.*	

Tab. 1. Bandwidth of upstream/downstream channels

The sideband bandwidth is totally independent on the video resolution and pixel clock. The max sustained data rate is 12 Mbit/s downstream and 9 Mbit/s upstream. That's enough to transferring and extending e.g. RS232, I2C or CAN. APIX serial sideband interface is asynchronously sampling the input signal (e.g. RS232) and transparently putting it through. At the receiver side the signal is provided together with its related clock signal (down_CLK, up_CLK).

3 Signal Quality & Signal Integrity

In-car display applications are extremely EMI-sensitive. Because Central Information Displays, Board Monitors, TFT-LCDs in Combo-Instruments, Head-Up Displays and Rear Seat Infotainment Systems feature a mix of resolutions and content, they may spread out a spectrum with a wide variety of different frequencies. The link distance (cable length) varies from 50 cm to 10 m and sometimes additional interconnections are needed to properly allow for partial pre-manufacturing, e.g. for rear seat monitors. To achieve interconnects to flat monitors, operating at lower ambient temperatures, small-scale cables at high flexibility are needed. Therefore, it is recommended that one ideal link be qualified to support all distances including interconnects, and all resolutions. Without matching the signal transitions to the cable impedance and its specific transmission characteristics (cables may have a widely differing capacitive load depending on their construction) it would not be possible to achieve long distance data transmission. And low EMI would be an unsolvable problem in terms of a cost-efficient application.

In this Eye Diagram, the measurement, taken at distance of 20 m, at receiver's input, shows the signal quality. Fig. 3a shows an Eye Diagram with the need for high amplitude to achieve a minimum eye opening (without any Pre-Emphasis), while Fig. 3b gives an ideal opened eye at already reduced signal swing. "Reduced swing" results in lower radiation, thus supporting lower shielding at lower cost.

By the use of a newly developed CML (Current Mode Logic) output driver stage with adjustable pre-emphasis and adjustable output drive current, combined with an unique line code (= dynamic scrambling of the video signal), the serial high speed signal is barely visible in the RF-spectrum. With the dynamic scrambling of the video signal, the spectrum of the serial signal is independent of the applied video signal. This feature is prevents fixed frequency spikes in the RF-spectrum when static images like navigation maps or control menus of a car PC are transmitted.

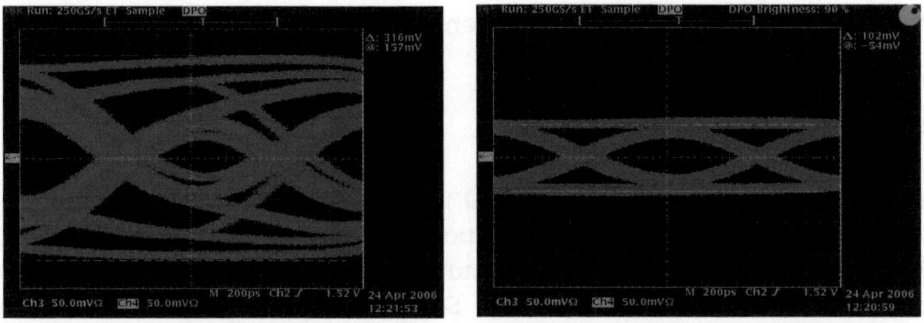

Fig. 3. without Pre-Emphasis (a, left) and with Pre-Emphasis and reduced swing (b, right)

3.1 APIX Evaluation Kit

To test all mentioned chip functions, Inova Semi-conductors offer a set of evaluation boards which perfectly support EMI- and Eye Diagram measurements. A dedicated slice of PC software allows adjustments to the entire chip and Link parameters, according to individual needs. An integrated video generator supplies different image frames for display timing tests. The set of boards comes with a ready-made 10 m LEONI-cable and automotive ROSENBERGER HSD connectors.

Fig. 4. APIX Evaluation Kit with automotive cable/connectors

4 APIX Enabling New Cost-Efficient Driver Assistance Systems

4.1 Display Links

Today, high-resolution displays (7 – 9″) are in common in all premium cars (central information displays, navigation displays, rear seat entertainment). Smaller displays (5″) are integrated into the car's dashboard cluster to permanently monitoring actual car settings. So it is mandatory to offer one generic link, satisfying all bandwidth demands thus simplifying the cabling. Presently, there are a number of programs that integrate the APIX chip into the display with just one four-wire interface to the entire display unit.

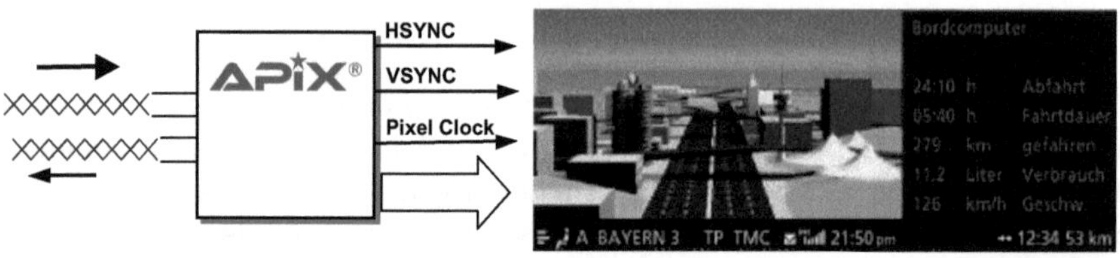

Fig. 5. APIX Rx - Direct Interfacing to displays

4.2 Camera Links

The ultimate goal, a self-driving car, is still a few decades away but, the first pre-requisites are already implemented in today's cars, such as the new CMOS-sensor based camera systems that enable to be proper detection of obstacles, signs, or manage proper lane alignment and keeping at speeds higher than 100 km/h in real time. When using the APIX, sensor head(s) do not require any image pre-processing by DSPs and therefore the camera unit can be kept to a very small foot print (2x2 cm - see Fig. 6). Therefore, the sensors can be mounted where needed with only two cable pairs required for video stream, control channel and power supply.

Fig. 6. Camera unit

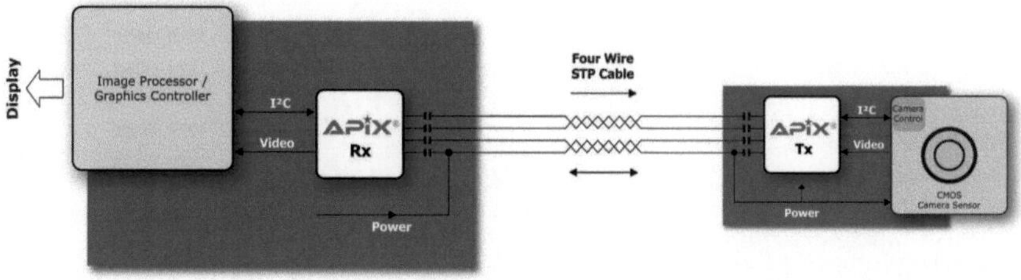

Fig. 7. Camera link using two cable pairs

References

[1] PD_INAP125, INOVA SEMICONDUCTORS APIX Data Sheet

[2] "Full video with one twisted Pair", Robert Kraus, Automotive Industries, Vol. 185/ issue 8/2006

[3] "Super schnell - Die neue Gigabit-Dimension im Automobil", Elektronik Journal, 05/2006

Axel Krepil
Inova Semiconductors GmbH
Grafinger Str. 26
81671 München
akrepil@inova-semiconductors.de

Keywords: APIX, LVDS, current mode logic (CML), GigaSTaR

The ConnectedDrive Context Server – flexible Software Architecture for a Context Aware Vehicle

St. Hoch, BMW Group Research and Technology
F. Althoff, BMW Group
G. Rigoll, Technical University of Munich

Abstract

Focus is pointed on the ConnectedDrive Context Server (CDCS), a central server component managing the situational context of an automotive human-machine-interaction, that is capable of providing key functionalities for a flexible prototyping process in the development and evaluation of "intelligent" vehicle behaviour. The main features are an object-oriented, shared knowledge management database and both generic, flexible I/O and knowledge analysis interfaces, offering a high compatibility for the connection to existing applications and for the implementation of intelligent reasoning algorithms. Our prototyping architecture is currently used in the development and evaluation of several context aware applications like a context-sensitive lane departure warning system (LDWS).

1 Introduction

Complex vehicles with certain cognitive capabilities have increasingly become incorporated in research areas of car manufacturers in various segments. Providing the driver with a maximum amount of security, luxury and comfort, future vehicles will be equipped with a wide range of intelligent systems like advanced driver assistance (ADAS) and in-vehicle information systems (IVIS). Research and development in these upcoming ADAS and IVIS has often been connected to the topic of contextual awareness and cognitive automobiles. Today, adaptive automotive system behaviour is mostly limited by the lack of context information and intelligent reasoning capabilities of the systems. However the recent evolvement of new sensor technologies like radar or lidar provides new possibilities in obtaining a more detailed view of the environmental situation and vehicle status. In addition to these "vehicle-related" information sources, "Driver Monitoring" sensors can yield aspects to one crucial, in today's systems left out component of the context: the driver himself. The knowledge of the complete situational context "vehicle, environment and driver" offers extended possibilities for the development of enhanced automotive applications, both ADAS and IVIS, capable of intelligent reasoning and adapting to the needs of the situational context.

At BMW Group Research and Technology, research in advanced vehicle technology has a long history. The central idea of the "ConnectedDrive" paradigm [1] is an intelligent, symbiotic combination of these contextual information fields of an automotive framework as illustrated in Fig. 1. The vehicle environment, the current driving state parameters and driver information can be interpreted as the vertices of the so-called automotive contextual triangle, representing an image of the global situation context. A cognitive vehicle would observe all these areas to offer an evolution of higher level context aware applications. One missing aspect, e.g. driver information, limits the intelligence of future ADAS and IVIS developments. Following this paradigm of situation sensitive vehicle behaviour, we have developed a new prototyping architecture, which provides us with a very flexible and powerful tool chain to develop and evaluate context aware vehicle applications, covering the three areas of contextual information in one single vehicle. In this work, the latest research of BMW Group Research and Technology in the field of intelligent vehicle behaviour is described in detail.

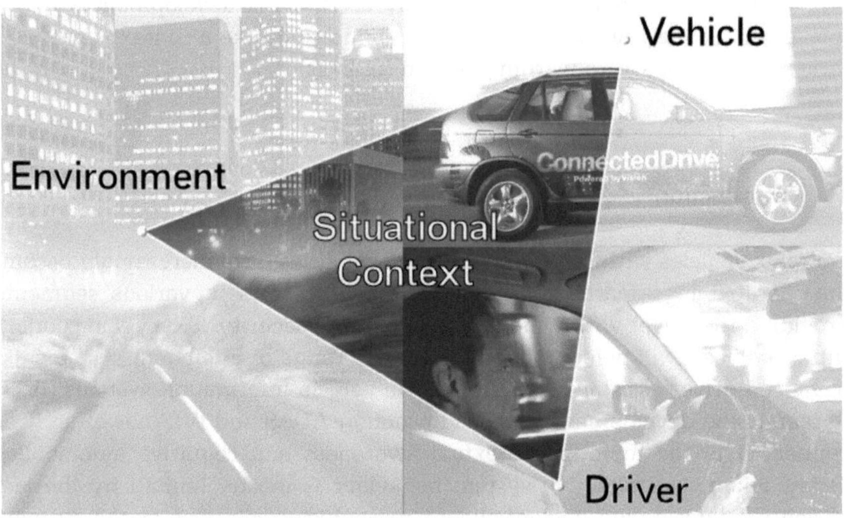

Fig. 1. Environment-Driver-Vehicle: The situational context

The following sections are organized in the following way. Section 2 gives a short overview of related work on context awareness in automotive research fields. The developed architecture is described in detail in section 3. Section 4 gives a short overview on our prototype vehicle and one example of a context aware application, that we developed using our architecture. A summary of conclusions and our future work is presented in Section five.

2 Related Work

Dealing with context awareness always demands a definition of the terms "context" and "context awareness" first, as their meaning has often been adjusted to the field of work it is used in. Dey et al. [2] have formulated a rather global definition proposing that context is "any information that can be used to characterize the situation of an entity. An entity is a person, place, or object that is considered relevant to the interaction between a use and an application, including the use and applications themselves". A context aware system "uses context to provide relevant information and/or services to the user, where relevancy depends on the user's task". We believe that this definition suits our research area very well as it can be adapted to the idea of the "ConnectedDrive" context with the core entities being the vertices of the context triangle shown in Fig. 1.

Research on context and context awareness has drawn much interest in recent years in both automotive and non-automotive areas, where it is most often summarized under the term of "ubiquitous computing". Automotive research work in this area can be separated in the development of context management architectures and context - aware applications. Several EU-funded projects have dealt with the idea of a context management component.

The Project COMUNICAR (Communication Multimedia Unit Inside Car) proposed an Information Manager (IM), which collects the feedback information of assistance-, telematics- and entertainment functions and estimates the driver's handicap respectively workload according to the impact of the current driving and environment situation. The overall system-feedback is adapted using a combination of the derived information parameters and the priority of the information display requests in order to enable a safe interaction between the driver and a multitude of in-vehicle systems [3].

The EU-Project AIDE (Adaptive Integrated Driver-vehicle InterfacE) is following a similar approach to maximize the efficiency and the safety benefits of advanced driver assistance systems while minimizing the level of workload and distraction imposed by in-vehicle information systems. The project architecture is based on a central high level perception instance, called the Driver-Vehicle-Environment (DVE), which senses and estimates the state of the driver, the vehicle and the environment. A communication assistant, which then controls a secure display of applications feedback information to the driver through the vehicle human-machine-interface [4], then adapts the applications feedback behaviour according to the contextual information provided by the DVE.

The NHTSA project SAVE-IT (Safety Vehicles Using Adaptive Interface Technologies) also has focussed on a central component monitoring the roadway conditions, the vehicle's and driver's state parameters [5] in order to develop, demonstrate, and evaluate the potential safety benefits of adaptive interface technologies that manage the information from various in-vehicle systems.

While the results and experiences of these projects have shown that there is a lot of potential in the idea of a central information processing unit, we think that context management is directly connected to the topic of context awareness and knowledge analysis, which in the development of automotive applications hast often been linked to the research topic of driver intention analysis, which several research projects have dealt with for adapting user interfaces and ADAS. Geiser and Nirschl [6] have suggested that the Driver Warning System (DWS) should analyse intention information for suppressing premature warnings in the European PROMETHEUS program. Likewise, in the United States the NHTSA Benefits Working Group [7] has proposed that an "ADAS like a collision avoidance system should be intelligent enough to discern driver's intention (e.g. intent to change lanes, lane change start) though this is difficult to do. If available, such indicators might selectively alter the drive alerts or warnings" (e.g. thresholds, presentation mode, stimulus magnitude, etc.) and reduce the number of nuisance alerts thus increasing the driver's acceptance of these assistance systems [8]. Regarding this development of intent inference algorithms, the NHTSA-Project SAVE-IT has recommended to use a combination of theory-driven and data-driven approaches including methods from the machine learning sector and implying the use of driver monitoring information [8,9].

Looking at the previous work on context management and awareness in the automotive area and motivated by the robustness and flexibility of inter-human communication, we see highly integrated contextual awareness as being a key component to provide an intuitive human-machine interface to the wide range of functions the driver is being confronted with in a modern driver's working place. We therefore have decided to merge the ideas of central shared knowledge management entity and intelligent reasoning capabilities with focus on a flexible framework offering both powerful knowledge and intention analysis services for the application development phase and a generic usage of derived knowledge parameters in the application runtime phase.

3 Architecture

A cognitive vehicle has to be aware of all available context information. By analyzing the time course of the context, it is able to interpret the situation and predict the necessary behaviour for upcoming situations. A basic system layout of a context aware vehicle application is shown in Fig. 2. It receives raw context information from a wide range of different sensors operating on various levels of abstraction. It processes this data, and adapts its behaviour or actions based on the results of a context inference engine, analysing the contextual situation.

This inference engine possesses on intelligent reasoning capabilities, which either can be realized by applying decision rules based on a-priori knowledge or by any other statistical learning scheme, e.g. neural networks or support vector machines, which learn their classification functions in an offline processing of labelled raw sensor-data. The inference engine then uses these models to combine incoming raw context parameters from the sensors and generate high-level knowledge context parameters, which cannot be measured by a sensor. Examples for this so-called meta context parameters are any kind of driver intention analysis results either related to the primary driver task, e.g. lane change intention, or any other secondary task, e.g. suggesting the destination in a navigation system based on an analysis of the entries in the calendar. This aspect of intelligent reasoning capabilities for the generation of context aware behaviour is a key constraint of our architecture.

One major drawback of the basic layout showed at Fig. 2, is that the knowledge generation process is application-bound, and therefore it is only used for the behaviour of the application with the results kept inside of it. Each application is producing its own application specific context resulting in a distributed knowledge or context management. In order to obtain a flexible platform we have decided against this type of system topology for our context aware prototyping architecture.

In order to separate the data pre-processing and intelligent reasoning capabilities from the application, we have chosen to build a central knowledge management component, the ConnectedDrive Context Server (CDCS). The principle layout of the architecture is shown in Fig. 3. The CDCS is the core component of a client-server-based knowledge management and distribution architecture.

Incoming raw sensor data is collected by a central I/O layer and forwarded to the CDCS. Raw contextual information is then transferred into an object oriented knowledge representation inside the core context management unit. Each object has its own value and timestamp history (memory functionality) and also a set of methods delivering statistical values like mean or derivates etc. and quality values.

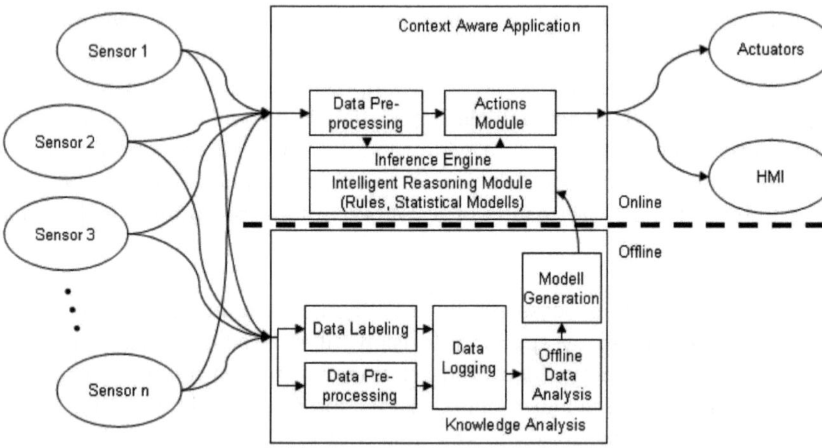

Fig. 2. Layout of a context aware application

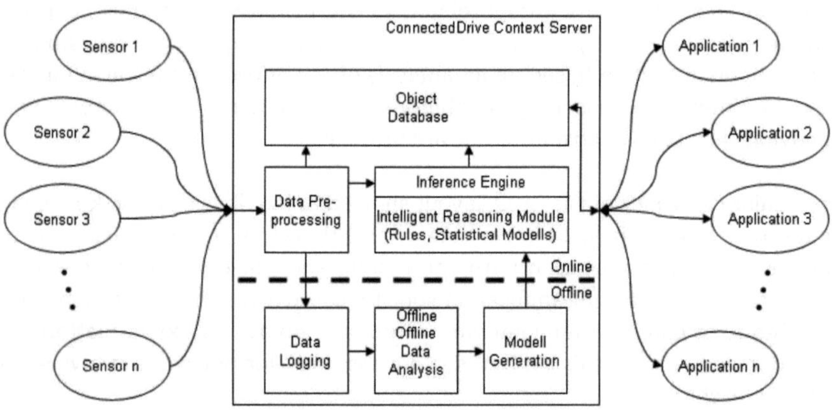

Fig. 3. ConnectedDrive context server architecture

Each parameter is assigned to a parameter management agent monitoring the objects state and when necessary updating the objects attributes.

The complete data I/O management of the CDCS is provided by its communication agents (CA). They listen for incoming request from external client applications, providing the information which context parameters an application would like to request and also which knowledge it might want to contribute to the universal shared context database. The submission requests of the applications have to match a defined XML syntax (see example request in (Fig. 4), defining the requested I/O operations. A CA then checks the compatibility of the request to the existing knowledge base and initializes the I/O interfaces according to the communication request. After that it prepares the object database for possible new context parameters from the application and opens an outgoing data stream providing the application with the requested context parameters. We have chosen the UDP protocol for incoming and outgoing data streams to secure the compatibility to a wide range of programming languages to use in our development process of context aware application.

```
<context_request application_name="app1" application_adress="127.0.0.1">
    <parameter_request udp_port="xxxxx">
        <parameter name="param_1"/>
        <parameter name="param_n"/>
    </parameter_request>

    <parameter_insert udp_por="yyyyy">
        <parameter name="param_x" parameter_type="Double"/>
    </parameter_insert>
</context_request>
```

Fig. 4. XML-based context request

The second key feature of the CDCS beside its data management and distribution function is its both flexible and powerful toolkit and interface to generate high-level information parameters, so-called meta parameters. These context values can not be measured directly, but can be calculated through an intelligent combination of raw context parameters. For example, the driver's intention could be inferred out of his steering and gaze behaviour. For this purpose the CDCS is equipped with a flexible interface for the integration of statistical models for knowledge generation processes like neural networks, support sector machines or other learning techniques known from the field of machine learning.

The interface is designed for models generated in an offline training and evaluation process, which is done with the "Waikato Environment for Knowledge Analysis" (WEKA) [10], a powerful tool offering a wide range of data mining and machine learning capabilities provided with a flexible JAVA API. Furthermore the CDCS not only offers the interface to integrate these models in an agent-based generation process for meta parameters in real-time, but also provides the toolkit to record, label and prepare the data needed for the training of the models. After the generation of a new model, it is inserted into the intelligent reasoning module of the CDCS, where it is bound to a so-called enhancement agent, which uses the model to create a new meta context object in the object database. In this way our architecture offers a uniform process for the development, evaluation and usage of intelligent reasoning capabilities. One example of generating a meta context parameter in the described process to enhance a ADAS with intelligent reasoning will be mentioned in Section 4.

Besides a central knowledge management server, our framework also envisions one central Human-Machine- Interface (HMI) management instance, which, using the knowledge of the CDCS, arranges all applications' requests for HMI output modalities in a queue in order to guarantee an optimal context sensitive display of the applications outputs. Details about this component will not be topic of this article. The following section deals with our prototype vehicle and the first example application, a context sensitive lane departure warning system, which has been enhanced with intelligent reasoning capabilities using the tool chain of our context-aware framework.

4 Prototype Vehicle and an Example Process Scenario

Former studies have shown that research projects dealing with context awareness benefit most from a prototype application that can be experienced in real-life situations. We therefore have built up an experimental vehicle, whose hard- and software has been designed to provide the best possible framework for the development and evaluation of context-aware applications. It has been equipped with a wide range of sensor techniques (e.g. a lane detection camera system, a radar sensor for vehicle detection, a driver monitoring camera system (Fig. 5) to acquire a large set of context information from the areas environment, vehicle and driver. We have installed our context management prototyping software platform on a standard Windows based personal computer and have connected it to the vehicle's CAN bus system to obtain the data from the sensor systems. In the following we want to briefly describe an example prototyping process for a context aware driver assistance system we have started to develop and evaluate using our architecture.

We have chosen a lane departure warning system (LDWS), an ADAS application, which is already available on the market, to evaluate the impact in intelligent system design by introducing enhanced cognitive vehicle capabilities. In our LDWS we have used CDCS architecture to adapt the application to the driver's intention. In case of departing the current lane, the system has to judge if this is intended or not. The criteria for a lane departure is fulfilled in case of the Time-to-Line-Crossing (TLC)-value [11] falling below a given threshold. The TLC-value indicates the remaining time until the lane marking will be crossed by the car's wheels under the assumption of a constant lateral velocity.

In our first experiments we wanted to use the architecture to create an inference model to estimate the driver's intention to leave his current lane by observing his steering behaviour and vehicle dynamics before the TLC value reaches its decision threshold. We therefore have implemented a basic LDWS, and used its system behaviour to trigger the recording of context data for an intention analysis. During several hours of driving on German highways, where the drivers have not been told about the motivation of the drives, we have recorded observations of natural driving characteristics with multiple intended lane departure scenarios like lane changes, corner cuttings and lane oscillation, yielding in one set of characteristic context parameters every time the TLC value dropped below a defined threshold value. While it has been rather easy to collect data of intended driver behaviour, we have found out that is hard to model unintended lane departures due to the lack of data available. In comparison to other researchers, we have voted against a provoked generation of unintended lane departure scenarios in our data recording sessions with test persons, as we believe these experiments are too dangerous.

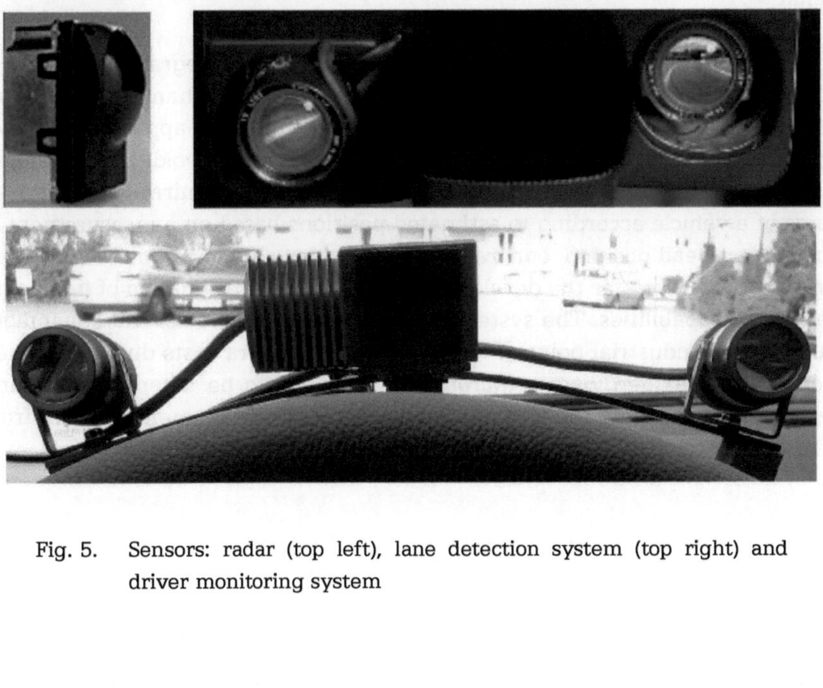

Fig. 5. Sensors: radar (top left), lane detection system (top right) and driver monitoring system

Instead we have used data from our researchers simulating unintended lane departures (with full vehicle and environmental security control) to get relevant data for our inference analysis. After recording and labelling the data in real-time, our CDCS architecture provides an interface to store the data already prepared for an offline data analysis with the knowledge analysis tool WEKA. We have had the possibility to run several test regarding the features' information theory statistics and the suitability and performance characteristics of several classification schemes. In this way we have calculated an optimal set of features and classification scheme for our inference problem. At this point of the design process of intelligent reasoning capabilities our architecture offers one of its most powerful features, providing an interface to directly integrate the models created with the WEKA environment into the inference engine of the CDCS, without writing a single line of programming code.

The key to this functionality is again based on the generic XML layout of the context management unit, which has been mentioned in section 3. Only a single entry has to be added to the central XML configuration file of the CDCS, causing the engine to instantiate a context enhancement agent. The agent begins to insert a new meta context parameter into the object database and then during runtime updates the inferred context parameter, in our case a lane departure intention, according to the input parameters defined in the provided classification module. The LDWS in a next step only has to request this intention value from the CDCS and adapt its warning behaviour to the calculated value.

This uniform process of consisting flexible creation and integration and powerful evaluation of intelligent reasoning capabilities for the enhancement of applications with adaptive behaviour can be used regardless of application type or scenario. Whether it is a different ADAS, like a collision avoidance system, or for example an adaptive application which modifies the mirror and seat settings of a vehicle according to estimated positions based on a measurement of the drivers head position, our framework provides a uniform and independent process framework for the development and technical evaluation of intelligent reasoning capabilities. The system architecture yields the several advantages but from an industrial point of view also involves extra costs due to the additional soft- and hardware requirements. This has to be taken into account, when further pushing the idea of a central context management instance from a research architecture to a component of a future series vehicle.

5 Conclusion and Future Work

In this work, we have presented a prototyping architecture that is capable of managing the situational context of the automotive human-machine-interaction. In comparison to other recent approaches to context management systems (AIDE, COMMUNICAR) our architecture has been designed to create a uniform research process for context aware behaviour in automotive applications. The both flexible and generic software framework offers great possibilities for a fast and uniform prototypical development, implementation and evaluation process. Furthermore, the flexible interface to a powerful data mining and machine learning tool has great potential for a detailed evaluation and comparison of a multitude of learning schemes for different intelligent reasoning tasks without time-consuming conversions between different toolkits. The server-client-based system layout enables us to easily connect existing applications to our framework providing them with either completely new sensor information or offering new enhanced context parameters for a more intelligent behaviour of these applications.

The work on the topic of contextual awareness in an automotive environment is strongly connected with multiple new sensor and machine learning technologies. It has been found that there are still some problems, which have to be solved in the related sub-topics. Sensors and recognition algorithms will have to be improved to provide more reliable context information to the inference algorithms. Furthermore, a systematic process not only for evaluation of the technical reliability of context sensitive algorithm design but also for the psychological perception of the adaptive behaviours by the driver has yet to be developed.

Nevertheless, our system has made a significant contribution to our research of advanced human-machine-interaction concepts and context sensitive applications. Our first experiments with an adaptive LDWS have shown that intelligent application behaviour based on an overall context analysis has clear potential to enhance the basic application regarding its situational awareness. We therefore plan to push forward the research and development of highly integrated contextual awareness as a key component of future automotive applications. We will explore the creation of completely new and the enhancement of existing ADAS systems, e.g. a frontal collision warning system. We further will evaluate the impact of enhanced situation-awareness on IVIS, for example in a dialog system equipped with reasoning capabilities based on affective computing aspects.

References

[1] T. Bachmann, K. Naab, G. Reichart and M. Schraut,"Enhancing Traffic Safety With BMW's Driver Assistance Approach Connected Drive", Torino, Intelligent Transportation Systems Conference, 2000

[2] Anind K. Dey and Gregory D. Abowd, "Towards a Better Understanding of Context and Context-Awareness", HUC '99: Proceedings of the 1st international symposium on Handheld and Ubiquitous Computing, Karlsruhe, 1999

[3] R.Schindhelm, Christhard Gekau & Marika Hoedemaker „COMUNICAR Information Manager: Ergebnisse der Felduntersuchungen", Garching, Tagung Aktive Sicherheit durch Fahrerassistenz , 2004

[4] H. Kussmann, H. Modler et. al, "System Architecture, data flow protocol definition and design and AIDE specifications", AIDE Deliverable D3.2.2, 2004

[5] J. Lee, M. Reyes et al., "Safety Vehicles using adaptive Interface Technology (Task 5) Final Report: Phase 1", 2004

[6] G. Geiser and G. Nirschl, "Towards a system architecture of driver's warning assistant". In Parkes, A.M. & Franzen, S. (Eds.) Driving future vehicles, London: Taylor & Francis (pp. 251-263), 1993

[7] NHTSA Benefits Working Group, "Preliminary Assessment of Crash Avoidance Systems Benefits", National Highway Traffic Safety Administration, Washington, DC, http://www. itsdocs.fhwa.dot.gov/ , 1996

[8] M. Smith and H. Zhang., "Safety Vehicles using adaptive Interface Technology (Task 8) A Literature Review of Intent Inference", 2004

[9] M. Smith and H. Zhang., "Safety Vehicles using adaptive Interface Technology (Task 9) A Literature Review of Safety Warning Countermeasures", 2004

[10] Ian H. Witten and Eibe Frank (2005) "Data Mining: Practical machine learning tools and techniques", 2nd Edition, Morgan Kaufmann, San Francisco, 2005

[11] D.A. Pomerleau, T. Jochem, et al., "Run-Off-Road Collision Avoidance Using IVHS Countermeasures", NHTSA Final Report, DOT HS 809 170, 1999

Stefan Hoch
Hanauerstrasse 46
80992 Munich
Germany
Stefan.Hoch@bmw.de

Dr. Frank Althoff
Knorrstrasse 147
80788 Munich
Germany
Frank.Althoff@bmw.de

Prof. Dr.-Ing. Gerhard Rigoll
Theresienstrasse 90
80333 Munich
Germany
Rigoll@tum.de

Keywords: context, context awareness, context sensitive applications, intelligent reasoning, machine learning

Prof. Dr.-Ing. Gerhard Rigoll

Components and Generic Sensor Technologies

Sensors for Active and Passive Safety Systems

T. Goernig, Continental Automotive Systems

Abstract

Passive safety systems have contributed very much in reducing injuries or fatalities related from vehicle accidents for the last decades. Passive restraint systems such as belts and airbags have reached an installation rate of almost 100%, providing an optimum of protection for the vehicle occupants. A further increase of road traffic safety requires also new generations of sensing systems. The systems shall also be capable to address the area of passive safety and active safety. With this type of sensing systems relevant data is collected in the surrounding of the vehicle. By combining this primary data with additional data from the vehicle, e.g. vehicle speed, steering angle, driver interaction etc. it is possible to activate non-reversible restraint devices for occupant protection as well as pedestrian protection systems. The system is also used to support braking functions to reduce the stopping distance of a vehicle. At low speeds a crash can most likely be avoided, even an automatic emergency braking would be possible with this type of systems in the future.

1 Safety System Development

Vehicle safety systems are standard equipment since the late 70's. Due to constantly increasing legal requirements like the FMVSS208 [1] or ECE regulations [2], the demands on vehicle safety systems also increased constantly. A strong market pressure due to independent tests from organisations like EuroNCAP [3] in Europe, IIHS [4] in the US also accelerated the introduction of the safety systems into the vehicles. The first restraints devices have just been lap belts, today there are at least three anchor automatic belts, supplemented by various airbag systems.

State of the art restraints systems consists of at least buckle or belt pretensioner, frontal and side airbags, head protection systems, battery and fuel cut off systems as well as the activation of an emergency or rescue calling system. In some luxury vehicles we also can see active steering columns, active pedals, active anti submarining systems, knee bags and active head rests, even on the rear seats of such vehicles.

All these systems are activated by pyrotechnical or electromagnetic actuators and have in common that the decision for activation of the restraint devices is based on the signals acquired from acceleration sensors during a crash. The sensors are part of the control unit that is installed at a central place inside the safety structure of the vehicle [5].

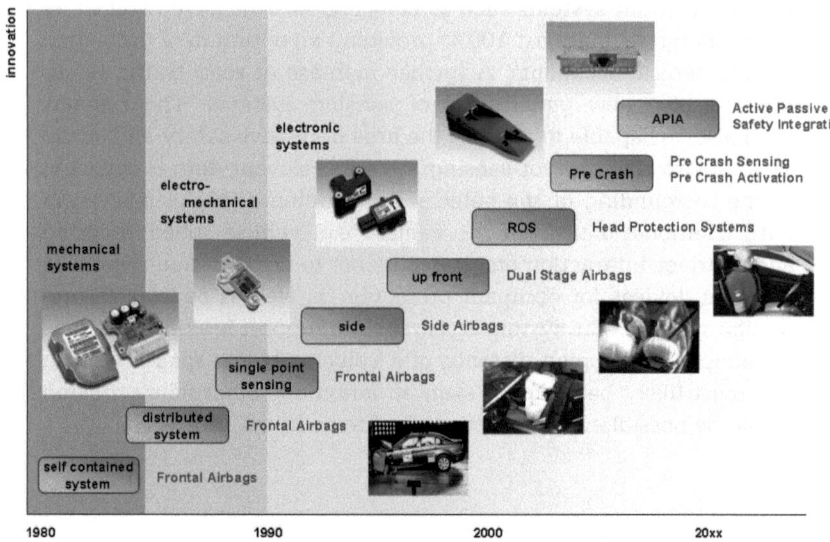

Fig. 1. Sensing and system development

The development of different types of restraint systems also required different type of crash sensors. State of the art sensing systems are using a number of accelerometers for the detection of a front impact, side impact or rollover event. Additionally pressure sensors are mounted inside the vehicle doors to get a fast and reliable detection of a side impact. The signals from the external sensors are transmitted to the central control module and are processed in an algorithm that decides if and what type of restraints devices are activated to provide the best protection of the occupants during a crash event.

2 Advanced System Requirements

For an advanced safety system it is required to get reliable information about a possible impact before the actual impact occurs. Depending on the range of the sensor that are used, the information is used for driver warning, for driver assistance and finally for occupant protection. To generate reliable information before the possible impact, it is required that the sensor measures the proximity of objects within the complete area of a vehicle. For passive safety applications the sensing range is defined to measure the relevant information 0.3 s, equivalent of a distance of 5 m at a vehicle speed of 50 kph, before the impact. The complete sensing area in front of the vehicle is 27° wide to also cover oblique, full front and offset impacts.

The relevant information is available to the system before the actual impact occurs. Based on this information it is known which occupant restraint device should be activated. It is also possible to estimate how severe the impact will be since the relative impact speed is calculated from the information available from the sensor. This information is used to control occupant restraint devices like adaptive airbag systems (e.g. dual stage airbags), pretensioners adaptive load limiters or also reversible restraints devices like motorised seat belts (MSB).

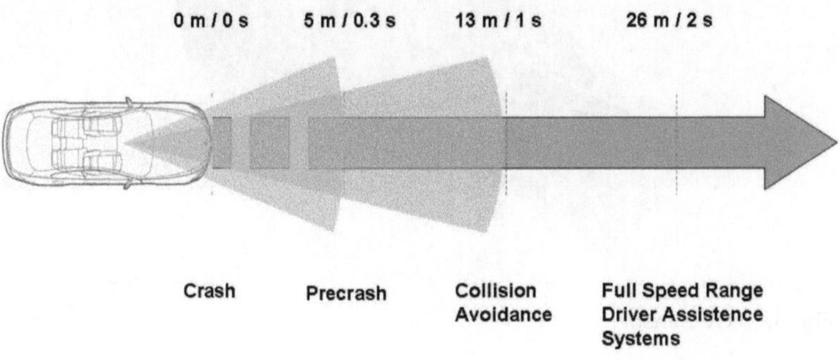

Fig. 2. Sensing ranges

Optical sensor systems have shown a very high potential to fulfil this ambitious requirements. They are also able to detect standing or moving objects, compared to systems that are base on other technologies. Optical sensors have a scalable resolution and can be built small enough to even fit in mini sized vehicles. Additional functions that are also measured with optical sensors, e.g. rain sensor, light and temperature sensing etc. can easily be integrated. Based on this arguments optical sensors have a high potential to become a real multifunctional low cost measurement system in the future.

3 CV Sensor

The pre-crash sensor (CV Sensor) developed by Continental Automotive Systems consists of an optical sensor system that is mounted behind the windscreen, in the upper area close to the rear view mirror. The sensor is primary used to measure the closing velocity of an approaching object. The measurement is based on the time of flight principle. A laser diode transmits a pulsed infra red light that is reflected by objects within the detection range of the sensor. Due to the installation position the sensor takes advantage of a clear view provided by the wipers when operated.

Picture: Temic

Fig. 3. CV Sensor

The integrated decoder calculates the runtime between the emitted and received laser pulse, which correlates with distance of the object in front of the vehicle. The sensor has three receiving channels which covers the complete vehicle front without gaps between the different channels. The three channels are used to detect the direction of the object and from repetitive measurements the differential speed of the object is calculated.

The sensor has been optimized to cover different functions. The sensor can be used for passive safety, active safety and pedestrian protection systems [6]. This is also reflected in the orientation of the sensor, relative to the horizontal line in the vehicle installation.

Parameter	*Description*
Principle	*Laser pulsed time of flight*
Wavelength	*905 nm*
Beams	*3*
Horizontal angle	*8°*
Vertical angle	*6°*
Distance Range	*≤ 10 m*
Distance Accuracy	*±10%*
Velocity Range	*5 - 150 kph*
Velocity Accuracy	*±10%*

Tab. 1. Optical sensor data

The primary application of the optical sensor is the usage as a pre crash sensor. The sensor system measures the distance of an object in the front of the vehicle. The speed of an approaching object is calculated from the distance information of repetitive measurements. With the wide beam and multiple receiver arrangement it also possible to calculate the direction of the approaching objects. Due to the fact that the relative impact speed can be calculated very precisely, the activation of restraint devices can also be staged much more efficiently.

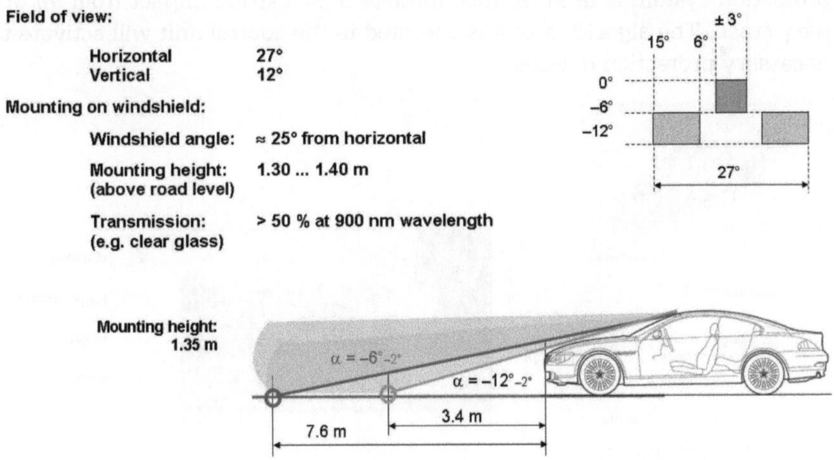

Fig. 4. Vehicle Installation

A further advantage of the sensor arrangement is that the information is independent from the vehicle structure, variations or ageing effects that might influence the crash performance of a conventional sensing system are eliminated.

4 Pedestrian Protection

A further application of the optical sensor system is active pedestrian protection systems. The introduction in Europe was decided from September 2005 on. The first systems include an active hood, which is lifted up for some inches to give additional deformation area, to reduce the physical load of the pedestrian. The sensors used for the first phase include fibre optical force sensors and acceleration sensors installed in the bumper. The next generation of pedestrian protection systems, that are introduced from 2010 on will include additional airbags at the a-pillar or the lower windscreen frame to protect the pedestrians from direct contacts with hard points of the vehicles. The purpose of the sensor in this application is to detect an object, in this case a pedestrian in the vicinity of the vehicle and to predict the impact time and speed of the object. A measurement window is opened to confirm the impact with an additional force sensor or accelerometers. The objects are tracked after the confirmation of the physical impact, "hard" and "soft" objects can also be discriminated by this method. Additional information from the occupant protection system is used for discriminate a pedestrian impact from an ordinary crash. The algorithm that is operated in the control unit will activate the necessary protection devices.

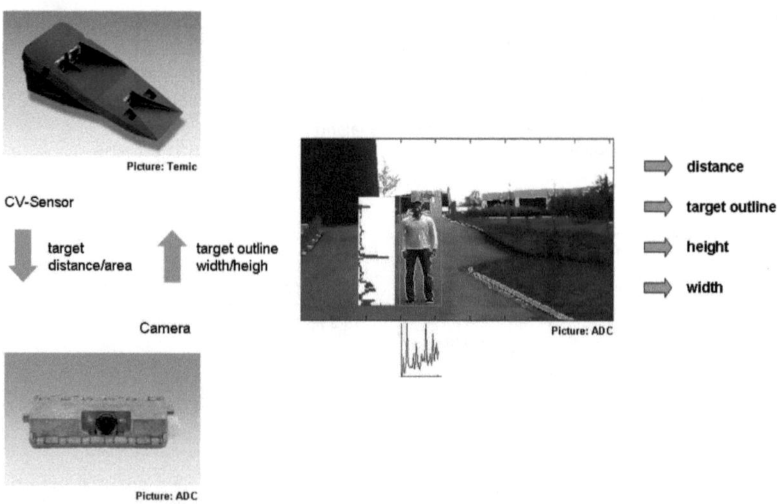

Fig. 5. Pedestrian classification system

Further demands will require sensor systems that are able to classify objects. The sensor systems will have to work with distance information and images that are taken in the close proximity of the vehicle front. The sensor systems will use high class image processing technologies to distinguish pedestrians, groups of pedestrians, cyclists etc. from other so called non vulnerable objects. This type of sensor system can be built by combining a high resolution 2D camera and a CV sensor. The sensor fusion will combine the high resolution image from the camera with the distance information from the CV sensor.

5 Active Passive Safety Integration

A further application that the optical sensor is designed for is the use in a system that integrates active and passive safety, as it has been shown in the Continental Automotive Systems APIA (Active Passive Integration Approach) project [7]. The information of the optical sensor is processed by also take additional information of the vehicle into consideration like the wheel speed sensors, steering angle, vehicle dynamics, brake and gas pedal activation. The primary data that is used for this calculation is the distance and velocity information form the optical pre crash sensor. Based on the information a danger potential calculated, allowing a controlled intervention in different phases:

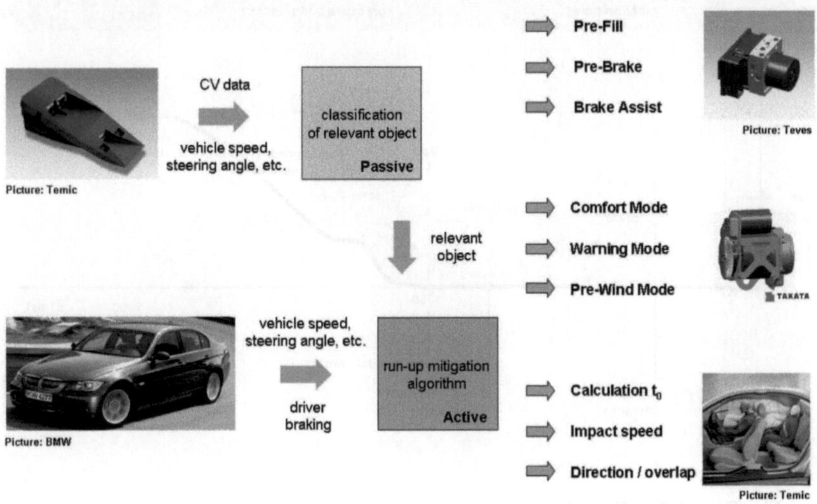

Fig. 6. Active passive safety integration

The pre-fill mode of the brakes is activated autonomously, eliminating the air gap between the brake pads and the brake discs in the first phase, in case of any approaching object that is detected in the driving path of the vehicle. The motorised seat belts are activated with a low force level, also as a feedback to the driver. The feedback is used to increase the attention of the driver to the possibly critical traffic situation.

The second phase, the danger mode, is activated in case that the driver releases the gas pedal. The brake system is activated with a maximum of 0.5g deceleration, already reducing the vehicle speed before the driver is able to push the brake pedal. This brake activation is equivalent to the braking action of an ACC system. The motorised seat belts are activated with a mid force level to secure the driver and the passenger in the seats.

In the last phase, the pre crash mode, the system autonomously calculates the necessary brake pressure that is required to stop the vehicle. When the driver is activating the brakes, the system supports the driver if he does not apply the required brake force, by increasing the brake force. In this case the system applies the maximum brake pressure.

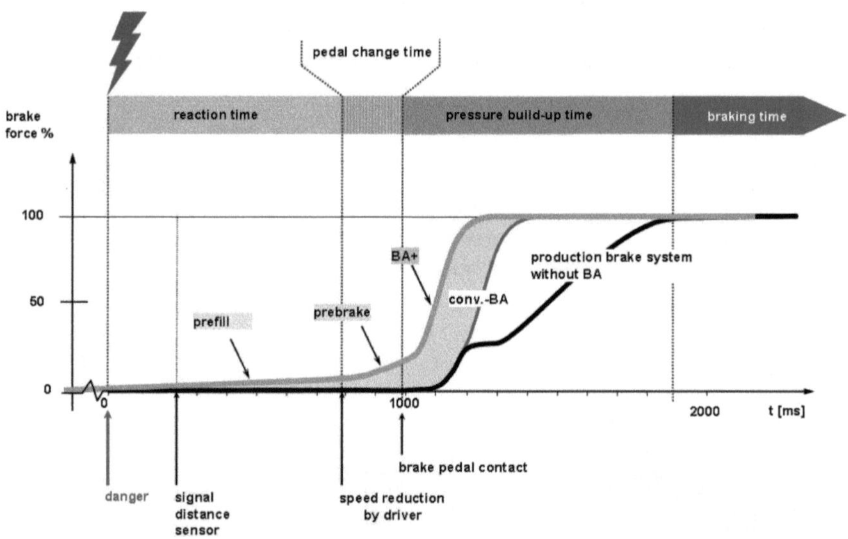

Fig. 7. APIA Functionality

The active passive safety system provides reliable protection and intervention up to a vehicle speed of 35 kph. With this system layout the majority of low speed frontal accidents are covered. Up to vehicle speed of 50 kph a significant reduction of the impact energy can be achieved.

Phase	*Mode*	*Action*
1	***Warning***	*pre-fill of brakes, motorised seat belt, low level*
2	***Danger***	*pre-brake, motorised seat belt, mid level*
3	***Pre crash***	*brake, motorised seat belt, high level*

Tab. 2. APIA Activation Modes

7 Safety System of Tomorrow

The trend towards integration of active and passive safety puts also new demands to the design of restraint systems. Another important issue is the wiring effort for this type of systems. It has to be reduced for costs, assembly, weight, and quality and reliability reasons.

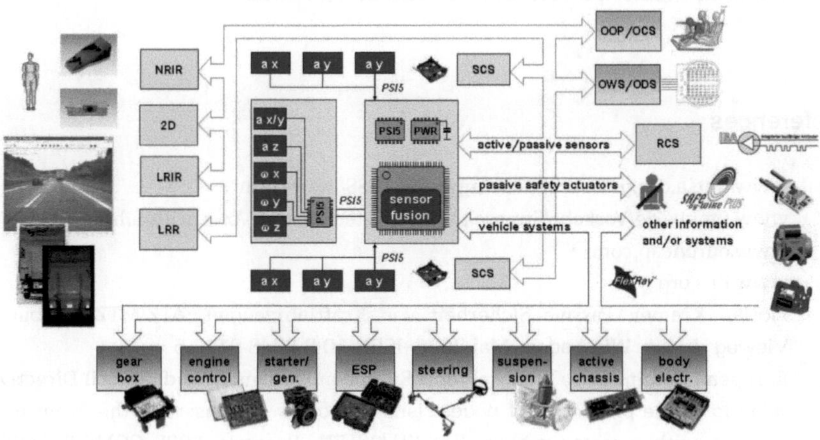

Fig. 8. Safety System of Tomorrow

Sensor clusters for active and passive safety functions will be used in future vehicles. The functions include ABS/ESC, vehicle stability, roll over prevention and all aspects of vehicle dynamics as well as passive safety. New types of bus systems will be used for the connection and transmission of information between the different modules. There will be different types of bus systems, depending on the different applications that we can find in future vehicle systems. There will be sensor bus systems as they are required for transmitting crash sensor information and there will also be sensor bus systems for smart peripheral devices such as buckle switches, seat track sensors or actuators like motorised seat belts. New standards will be introduced in future vehicles, PSI5 [8] will be used for fast transmission of sensor signals to restraint system controllers, Safe by Wire plus [9] is intended to be used for fast and secure bidirectional data transmission for more complex sensor and actuator arrangements. The activation of non-reversible systems is possible by using the potential of a sensor fusion system. By combining a distance measuring system and a vision based sensor, a much higher reliability of the information can be achieved, compared to the information derived from a single sensor.

Future restraint systems require new types of sensors and also new technologies. They will require pre crash sensors with different sensing ranges, resolutions, wide and narrow focused beams. There will also be no single sensor arrangement, to achieve an acceptable reliability of the signals, sensor fusions will be introduced. Pre crash sensors will also help to increase safety for all road traffic participants significantly. The challenge for the future is to find the best possible combination of the different available sensors and technologies to get a maximum of performance at economical costs.

References

[1] www.nhtsa.dot.gov/cars/rules/import/FMVSS/index.html

[2] www.bmvbs.de/Verkehr/Strasse-,1446/KfZ-technische-Vorschriften.htm

[3] www.euroncap.com

[4] www.iihs.org

[5] Florian Kramer, Passive Sicherheit von Kraftfahrzeugen, ATZ/MTZ-Fachbuch, Vieweg Verlag, Wiesbaden, Mai 2006, ISBN-10 3-8348-0113-5

[6] European Commission, Proposal for a European Parliament and Council Directive relating to the protection of pedestrians and other road users in the event of a collision with a motor vehicle, III/5021/96 EN, Buessels, 1996 COM(2001)389, July 11th 2001, ACEA Commitment relating to the protection of pedestrians and cyclists europa.eu.int.

[7] www.conti-online.com

[8] PSI5 system specification, www.PSI5.org

[9] Safe by Wire plus specification, www.PSI5.org

Thomas Goernig
Conti Temic microelectronic GmbH
Ringlerstr. 17
85075 Ingolstadt
Germany
thomas.goernig@contiautomotive.com

Keywords: ABS, ACC, ADC, APIA, CV Sensor, ECE, ESC, FMVSS, pre brake, pre fill, pre crash, PSI5, safe by wire plus

Failure Mechanism Analysis as Enabler for improved Test and Reliability Strategy - the Road to Success?

A. Rekofsky, C. Götte, A. Schingale, Siemens VDO Automotive

Abstract

The controlling units in safety, powertrain, comfort and networked vehicle application are speeding up to more and more complex systems. These systems have to survive long-term operation in a harsh environment, where humidity, vibration and, over all, high temperatures will impact reliability. On top of that, temperature and lifetime requirements are increasing, while materials, components and processes are still limited to their environmental conditions. Additional the market asks for zero ppm failure rates, extended warranty and decreasing development time. More reliable electronic systems with high integrated functionality within a shorter period of development time, new methods/models for reliability of components and materials and lifetime prediction are necessary. Understanding about failure mechanism is one essential topic for accelerated test methods, end of life prediction and decreasing testing time. A promising vision is the decreasing of development time using virtual qualification.

1 Introduction

Mobility, one of our main key for improvement is rapidly growing as main factor for advantage in competition. The base of mobility nowadays is the car. The impact of the automotive industry is also reflected by the huge effort inside research and development for strengthen innovation. Mainly the areas of safety, powertrain, comfort and networked vehicle are speeding up to complex and electronically controlled systems. Additional the permanent growing requirements ask for complex systems consisting of electronics, mechanics and software.

The introduction of this approach was enabled by highest system integration and design optimisation to the specific applications. These systems have to survive long-term operation in a harsh environment, where humidity, vibration and, most of all, high temperatures will impact reliability. On top of that, temperature and lifetime requirements are increasing. At the same time materials, components and processes are limited regarding environmental conditions and reliability.

As the market for future generation will require zero ppm failure rates and extended warranty, there is a substantial challenge for industry to provide technologically innovative and cost-efficient solutions. Additionally the automotive industry also calls for decreasing development time of automotive electronics. The current development time for electronic control units is 2-3 years for example. One task for the long period is the qualification time, by today it takes about 6 months.

2 Increasing Requirements

One trend in the automotive industry is the increasing of functionality which means higher integration caused of additional electronic components (e.g. sensors, actuators). Contrary to that the substrate area is continuously decreasing. And further on a comprehensible ask from our customers, for complete tested modules, implies the need of direct fixing control units on engine. On engine mounting reduces interfaces and leads therefore to higher reliability but parallel we have to withstand harsher ambient conditions. Fig. 1 shows the operating environment depending on mounting location.

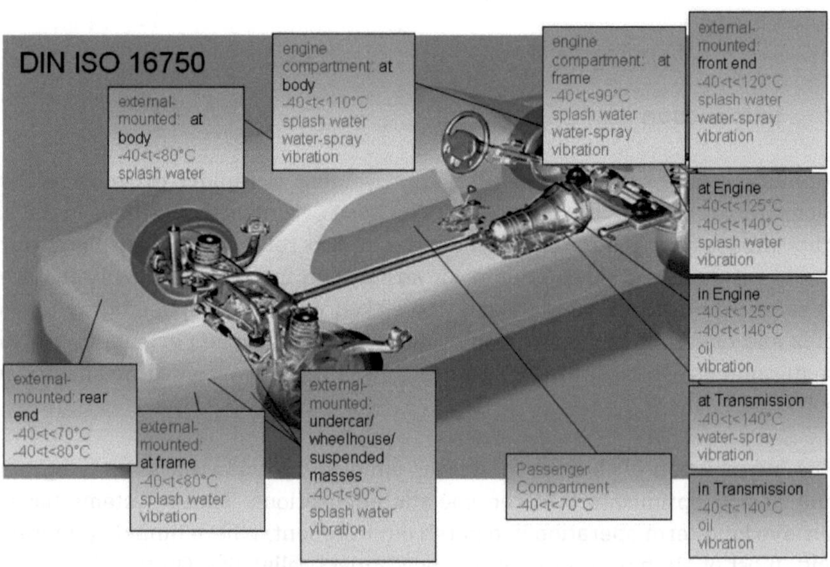

Fig. 1. Operating conditions depending on mounting locations

On top of that the development of novel power train concepts (e.g. power electronic which is essential for hybrid electronic vehicles) requires electronics for higher power losses. Therefore a better power supply, improved thermal management, suitable materials and advanced cooling systems are needed.

These harsher environmental conditions combined with longer warranty periods assume an extended product qualification, leading to longer testing times.

In contradictory to this trend is the request for reduced time to market. In 1985 the time to market was about 4-5 years; today it takes about 2-3 years. Each sample from development phase (A-, B-, and C-Sample) including redesigns needs a validation. Summarized this leads to a test time of more than half a year.

3 Field vs. Lab

The main goal of environmental tests in the lab is the reproduction of the environmental impact on control units in a car. The more you can accelerate these tests in the lab, the more you can shorten the development time and so reduce the time to market.

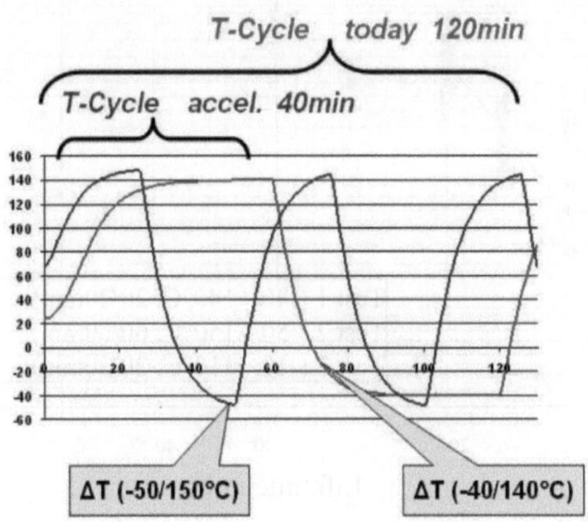

Fig. 2. Accelerated testing for mechanical cluster using Coffin Manson model

One example for a failure mode cluster and the features of acceleration are described by the example beyond. This failure mode cluster includes the interconnect technology like Al-bond wires and module functions like leakage, lamination of flex foil and welding points. As for these technologies the main failure model is just thermo mechanical stress an acceleration could be reached by increasing stress per time unit, cut the temperature storage (Arrhenius failure model) component out of the thermal curve and increase the gradient without exceeding the material limits (Fig. 2).

To get the Coffin Manson factors m for these functions at least two end of life tests with reduced storage time (blue curve) and different temperature ranges are needed. The following equations shows the Coffin Manson law with A, the acceleration factor, ΔT_t the test temperature range, ΔT_u the use temperature range and, m the Coffin Manson factor:

$$A = \left(\frac{\Delta T_t}{\Delta T_u} \right)^m \qquad (1)$$

Measurements of temperature profiles in transmission oil have shown, that the temperature in mounting location is a composition of macro cycles with low gradient and high temperature ranges, which corresponds to engine heating during one drive, and of micro cycles with high gradients and low temperature ranges corresponding to single power loss peaks during the drive.

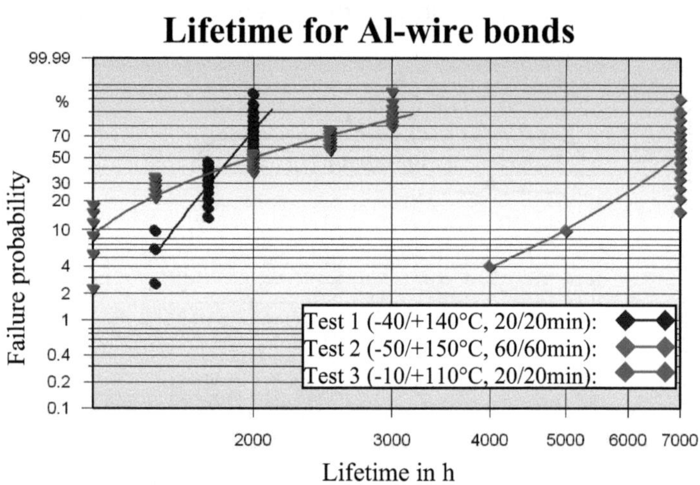

Fig. 3. Weibull evaluation for Al-bonds

To get the correlation of the today's lab test, which represents only to the macro cycles in field, to the accelerated tests without Arrhenius component, which represents the micro cycles in field, a third end of life test, with today's lab parameters is necessary. The evaluation with Weibull (Fig. 3) shows different gradients for the failure probability in the three tests. But the theory of Weibull says, that different gradients indicate different failure mechanism. These effect conflicts with the key idea of accelerated testing (produce the same failure mechanism in the lab as in the field, only in a shorter time).

The accelerated testing with test time reduction down to 67% by leaving out Arrhenius components has shown that a comparison between the different test methods is not possible and therefore an accelerating of today's standard test is not possible.

4 Physics of Failure

In the field of automotive electronics, a fast and precise methodology to predict lifetime of electronic modules is required. Specified below is a more analytical concept to form a lifetime prediction based on physics of failure approach.

A typical control unit for automotive applications consists of three main types of elements: substrate, component and its interconnection, mainly solder attach. Each of these single element types shows characteristic failures. Goal of our work is the knowledge of the driving force in a physics of failure approach, which lead to an element-specific failure.

In a first step all electrical, thermal and mechanical effects on each element type have to be known. Second step is the investigation of possible interactions between each element type.

All theoretical work is supplied by experimental setups. Combining theoretical results with real stress parameters leads to failure models as basis for a lifetime prediction.

This approach will allow a precise lifetime prediction when a device under special load conditions will fail.

5 Virtual Qualification

To fulfil these opposite requirements – development of more reliable electronic systems with high integrated functionality within a shorter period of development time – research of new methods and models for reliability prediction of components, materials and product lifetime are necessary. A possibility to handle reliability and end of life prediction is a "life time prediction tool".

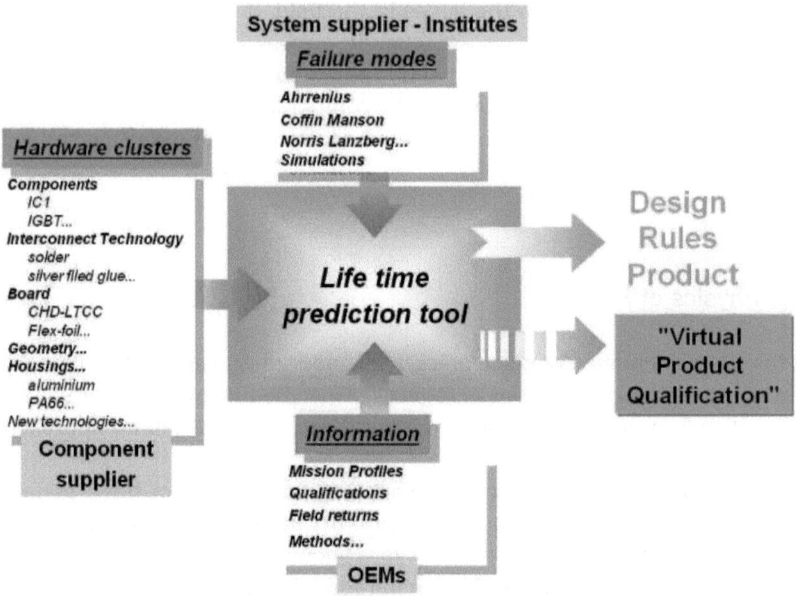

Fig. 4. Life time prediction tool

Out of all this previous considerations an over all merge in information seems to be the right way. The knowledge of all this useful and promising fields should give us the opportunity for an improved virtual simulation approach. Simulation should benefit from the results of the different methodical investigation and come to a suitable and mainly trustable performance. Qualification and test methods (lab, field etc.) can be adapted due to predicted entrapments during virtual qualification. This will reduce testing time to its basic needs. Additional this appreciation gives us detailed design rules for product and additional, where needed accelerated tests for failure mode cluster are available. It also enables a life time prediction for a designed product.

One challenge will be the question how to find the architecture for such a revolutionary way. The advantages are indisputable, but will there be an acceptance on the market. How can we find the boundaries for the requirements and how can we create the essential database and the system behind? In any case nobody could go this way by one's own. The whole supplier chains with the OEMs on the top in cooperation with institutes, all have to find the solution together.

References

[1] D. Wolf, Approaching Reliability for High Temperature Electronics for Automotive Applications, Paris 2005, The International Conference on High Temperature Electronics HITEN'2005

Andreas Rekofsky, Carsten Götte, Angelika Schingale
Siemens AG
Siemens VDO Automotive
SV P ED T IC
Siemens Str. 12
93055 Regensburg
Germany
andreas.rekofsky@siemens.com
carsten.goette@siemens.com
angelika.schingale@siemens.com

Keywords: accelerating, automotive, failure mechanism, lifetime prediction, powertrain, reliability, virtual qualification, vision

References

Intelligent Low-Power Management and Concepts for Battery-less Direct Tire Pressure Monitoring Systems (TPMS)

T. Lange, M. Löhndorf, Infineon Technologies AG
T. Kvisterøy, Infineon Technologies SensoNor AS

Abstract

Intelligent concepts to reduce the current consumption of direct battery based TMPS are mandatory in order to fulfil the OEM requirements for lifetime, operating temperature range and battery-size/weight.

In this paper an overview of standard TMPS applications and developments towards a reduced current consumption through intelligent low-power management will be presented. Overall goal is the reduction of the battery size and in a second phase the replacement of primary batteries by an energy harvesting concept. Therefore, different vibration harvesting concepts have been evaluated and will be discussed in view of possible TPMS applications.

1 Introduction - State of the Art TPMS

The most common TPM systems worldwide are battery-based systems. A typical configuration is represented by a micro-machined pressure sensor, a micro controller for processing, a radiofrequency (RF) transmitter to transmit the data to a central receiving unit and a battery as power source. Fig. 1 shows a schematic structure of a direct TMPS module.

For the RF-Transmission from the wheel to the Electronic Central Unit (ECU) the Industrial Scientific Medical (ISM) frequency bands are used (in Europe 434 MHz / 868 MHz and in the USA 315 MHz / 915 MHz).

The wheel TMPS module from Fig. 1 is fixed in an airtight housing by a molding process except the pressure hole. Due to high acceleration forces in the wheel a safe installation of the module on the rim is required. Therefore, the modules will be fixed direct on the valve or with a rotary belt on the rim.

Fig. 1. Schematic of a battery based direct TPMS module

Fig. 2 shows the main electrical components of a state-of-the-art TPMS module. Optional components are represented by a dotted line.

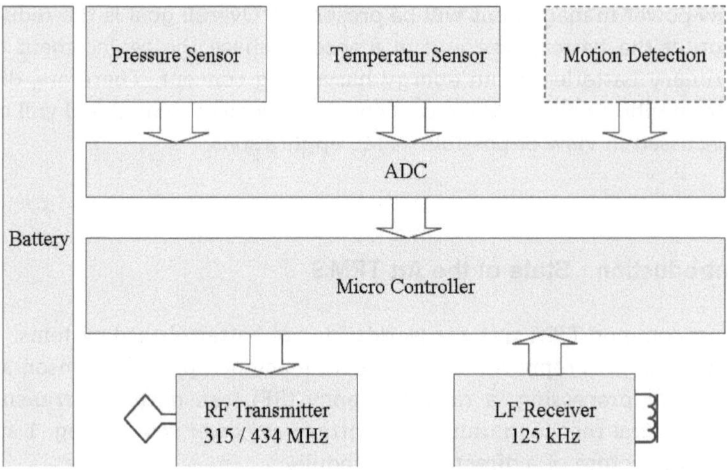

Fig. 2. Electrical components of a battery-based TPMS module

An additional acceleration sensing device is used for motion detection and enables the possibility to decide whether the car is parking or driving. Based on this information, the microcontroller will manage how often the pressure and temperature measurement is performed and transmitted via the RF link. This effective energy management allows to maximize the life of the battery as well as to minimize the current consumption. [1, 2]

2 Requirements

Inputs of technical requirements for Tire Pressure Monitoring Systems are coming from different sources. The US National Highway Traffic and Safety Administration (NHTSA) released the Federal Motor Vehicle Safety Standard 138 [3] to warn the driver, when a tire is significantly under inflated. Therefore, the vehicle manufacturer requires typical technical features in their product specifications to fulfill the legal requirements for the US and the safety demands for other countries of the world. In March 2006 an international standard was published (ISO/DIS 21750) [4]. Tab. 1 gives an overview about the key requirements from the different sources.

Feature	*NHTSA*	*ISO*	*OEM*
Pressure Warning	*25% under recommended inflation pressure*	*TBD*	*10% under recommended inflation pressure*
Response Time	*20 min*	*3 min*	*1 min*
Min. Operate Speed	*50 km/h*	*25 km/h*	*20 km/h*
Module Life	-	*6yr / 100000 km*	*10yr / 150000 km*
Malfunction Warning	*Yes*	*Yes*	*Yes*
Temperature Range	-	*-40°C ±85°C*	*-40°C ±125°C*

Tab. 1. Summary of regulations and standards

The summarized requirements in Tab. 1 have a direct impact of the current consumption of the TPM system. The next paragraph will describe the current consumption in more detail and outline methods for minimization of current consumption.

3 Current Consumption

As mentioned earlier, for a battery based TPM system the current consumption is the most critical issue. Finally, the current consumption determinates the size, weight and cost of the battery and therefore a significant part of the overall cost of the wheel module. An analysis of the current consumption will help to understand the split between different current consumers within a

TPMS IC. Fig. 3 represents the shares of the energy consumption over lifetime. The transmission power level is referenced to 5 db_m.

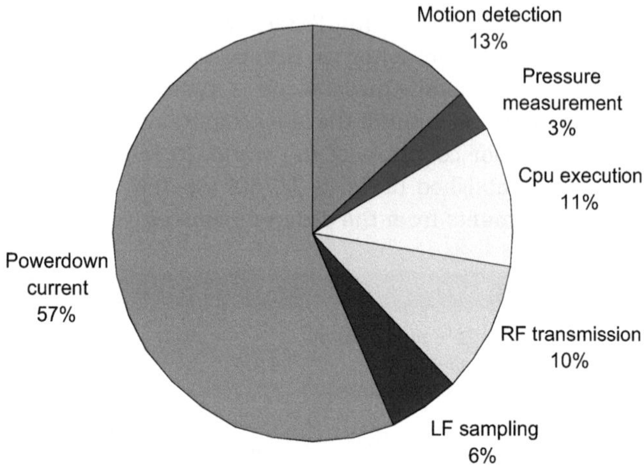

Fig. 3. Share of the energy consumption over lifetime of a TPMS module

The prorated biggest current draws are from RF transmission, CPU execution, motion detection and power-down current (leakage currents). A more detailed analysis of these specific modes will be discussed in the following.

3.1 RF Transmission

The RF transmitter modulates the data provided by the microchip, amplifies the signal and sends it out via the RF antenna. The trend goes to the RF integration into the TPMS chip, which allows smaller, lighter wheel modules and higher cost efficiency.

By utilizing higher data rates and therefore shorter transmission times the current consumption of the RF transmission can be significantly reduced. Another option for power saving could be the definition of so-called intelligent datagram. Tab. 2 shows a typical datagram of a current TPMS module.

Byte	*Description*	*Comments*
1	*Synchronisation*	*Synchronisation bytes for the receiver*
2	*Synchronisation*	
3	*Identification ID3*	*Unique 32 bit ID number*
4	*Identification ID2*	
5	*Identification ID1*	
6	*Identification ID0*	
7	*Pressure*	*Pressure Value*
8	*Temperature*	*Temperature Value*
9	*Diagnostics*	*Status Information*
10	*Check sum*	*CRC*
11	*End of message*	*1-2 bits*

Tab. 2. Typical Datagram of current TPMS module

The basic idea of intelligent datagram is to transmit only the ID as long as the pressure difference is zero. Thus, the pressure has to be measured on a regular basis and to be compared with the stored previous valued. Shows the result of such a short CPU routine no difference between the measured values the datagram can be reduced to the minimum. Although the duration time of the application program is increased and as consequence the CPU execution current, this approach saves to transmit a minimum of three bytes and thus means round about 25% of energy.

3.2 CPU Execution

About 11% of the total charge consumption is related to CPU execution. In order to minimize the cycle time a fast processor is required. Therefore, a single instruction cycle CPU is qualified. By a further optimization of the application program for example intelligent choices for equivalent operation functions, would reduce the necessary cycle time and in turn the current consumption. Thus, an optimization of the software according to the used hardware parameters could significantly reduce the overall current consumption and is inevitable for a lifetime requirement of 10 years as specified by most OEM´s.

3.3 Motion Detection

Although the detection of the vehicle motion costs a lot of energy due to the measurement of the accelerometer, it allows at the same time a more effective power management. Due to the fact that the vehicle is parked more than 90% on average over the 10 year lifetime the determination of the different modes (parking and driving) is of great importance since the number of measurements and RF transmission can be reduced to a minimum during parking mode. A special current saving mode so-called powerdown mode is used for the remaining time. Here, the current consumption is mainly due to leakage currents within the ASIC.

3.4 Powerdown Current

Due to the fact, that the TPM sensor is 85% of the time in powerdown mode this mode has also the largest current consumption with a share of approx. 50%. In this mode only a few parts of the TPM sensor are active the rest is switched off. Fig. 4 shows a block diagram of the active (marked green) and inactive (marked red) peripherals.

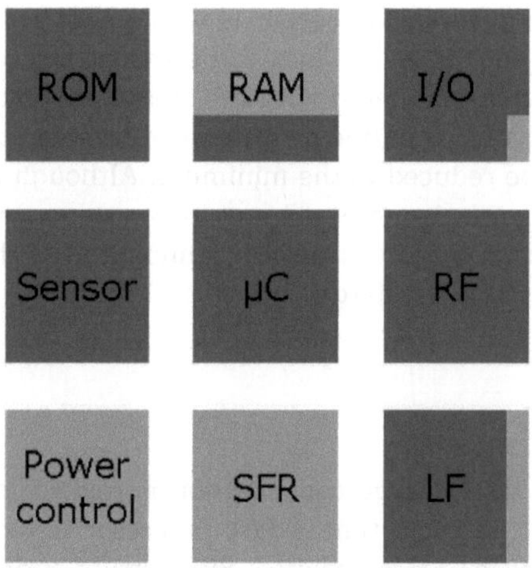

Fig. 4. Active and inactive parts of a TPM sensor during powerdown mode

The current draw can be mainly attributed to leakage currents within the ASIC. Therefore, several developments and technologies are explored to reduce the leakage currents. However, in contrast to the development to discrete components on the system IC level compromises between different applications, functions (flash memory, RF, etc.) in view of possible technologies have to be made.

A critical issue is the consideration of the powerdown current over temperature. In the high temperature range the current increase dramatically. The reason therefore is the leakage current, which are caused to the CMOS technology. The strategy to improve this situation is the usage of power switches, to cut off the power supply from needless circuits. Another solution could be to switch from bulk wafer to silicon on insulator wafer. The key advantage of this so-called SOI wafer technology is the improved isolation of the SOI area from the substrate.[5]

4 Power Sources for Tire Pressure Monitoring Systems

Primary batteries are widely used as power source. In addition inductive-coupled power schemes have been developed, but have not succeeded to replace batteries due to higher overall system costs. For future TMPS applications the development of novel power sources is of great importance in order to reduce the overall weight, seize and cost of the systems as well as to overcome the limited lifetime due to the provided battery capacity.

4.1 Li- Ion Primary Batteries

The majority of the commercial available direct TPM systems are using lithium-based primary batteries. Lithium (Li)- ion coin cells exhibit an excellent energy- density vs. weight ratio with an open- circuit voltage (V_{oc}) of about 3-3.5 V as well as an extended operating temperature range. In general three types of Li- ion primary cells are available using different cathode materials: polycarbonate monofluoride (BR type), manganese dioxide (CR type) and thionyl chloride (ER type). Fig. 5 shows a schematic cross-section of a Li- Ion primary coin cell. Usually, a polypropylene foil is used to separate the anode (lithium) from the cathode material. In addition an electrolyte is used to provide the ion exchange within the coin cell.

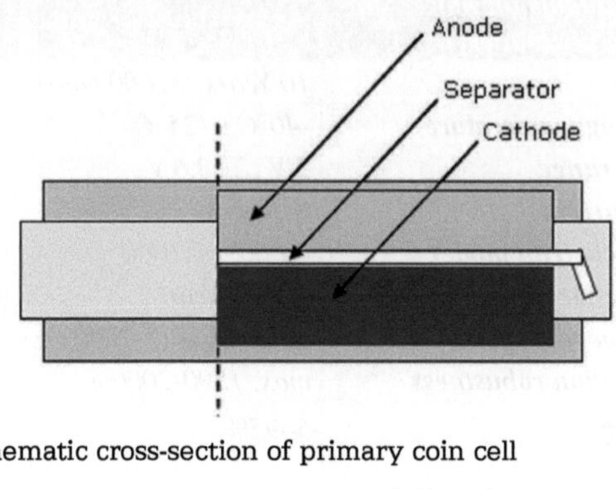

Fig. 5. Schematic cross-section of primary coin cell

Each cathode material has its own properties and advantages or disadvantages with respect to operating temperature, durability and fabrication costs. All three types have been used for TMPS applications, however due to superior high temperature performance, better durability and lower weight the BR and CR types are more often used. Lithium– thionyl chloride is a liquid cathode material under ambient conditions and show better low temperature behaviour due to the enhanced ion mobility, but needs a more complicated and expensive hermetic sealing. At low temperatures the internal resistance of the battery increases and the closed circuit voltage V_{cc} decreases leading to a drop of the supplied current. A few basic requirements for TPMS power supplies are listed in Tab. 3.

No coin cell battery could provide enough energy to power the TPM system continuously for a lifetime of 10 year. Therefore, the needed battery capacity strongly depends on the application program, temperature profile, transmission power level and ASIC specifications (Note: The average driving time of a privately used light vehicle can be estimated to be less than 5% of the total 10 Year lifetime). Usually coin cells of the type 2450 (24 mm diameter and 5 mm height) with a battery capacity of about 450-600 mAh are utilized within commercial TPM Systems in order to fulfil the requirements.

For primary Li- ion coin cells the capacity is mainly dependent upon the active volume or surface area of the batteries and thus the capacity decreases almost linear with decreasing size or volume. New material combinations have been developed to increase the energy density as well as new fabrication processes by thin-film deposition techniques. However, none of the reported research and development results could sustain all requirements for TPMS applications. Major drawbacks are the limitations within the operating temperature range and mechanical robustness.[6]

Requirements for power supply (battery)	
Lifetime	*10 Years "87600 hours*
Operating temperature	*-40°C ±125°C*
Voltage range	*2 V….. 3.6 V*
Pulse current	*8-10 mA*
Duty cycle (run mode)	*1:500 – 1:1000*
Self-discharge	*≤ 1% / Year*
Vibration robustness	*continuous (5-2000 Hz)*
Acceleration robustness	*max. 1500-2000 g*
Humidity	*5–95%*

Tab. 3. Requirements for TPMS power supplies

4.2 Energy Harvesting

Within the past years a vast number of research activities have been started in area of wireless sensor networks. Since the proposed networks comprises of thousands of individual sensor nodes which communicates with the nearest neighbours via wireless transmission small, efficient and inexpensive power sources are needed. In addition the reduction in size and power consumption of active electronic components as well as the advances in wireless communications enables the development of novel stand alone or mobile applications. However, the power source was, is and will remain the most critical issue for further product developments.

Environmental power sources such as solar-, wind-,water energy have already been used in macro scale devices. For MEMS (Micro- Electro- Mechanical-System) and/or CMOS based devices such as micro sensors new concepts have been proposed due to the reduction of the potential volume/surface of the generator. Possible energy sources for micro scale devices are: pressure-, vibration-, thermal gradient changes and micro- solar and fuel cells. In contrast to wireless network applications in the TPMS application the tire/wheel assembly should provide significantly more vibrations at higher amplitude. Thus, a vibration energy harvester might deliver enough average power to supply a TMPS device.

4.3 Vibration Energy Harvesting Concepts

Using mechanical vibrations as energy source is of great interest for industrial and automotive applications, since engines and turbines are inherent sources of vibrations. However, the frequency spectrum and amplitudes of these vibrations are very different and might vary during operation. Thus, an efficient vibration generator must be custom designed for the target application. In general, three different physical transduction mechanisms have been evaluated and published in the literature: 1. piezoelectric, 2. electromagnetic (inductive) and 3. electrostatic (capacitive). All three mechanisms have been widely used for inertial and pressure sensors as well as for actuators. The basic principle of all mechanical-to-electrical energy converters is shown schematically in Fig. 6. The spring suspended inertia mass will start to translate (oscillate) if the whole system is externally excited. Kinetic (mechanical) energy is converted into electrical energy by the induced electrical damping. Losses are implemented by the mechanical damping. The more induced electrical damping the merrier the converter efficiency.

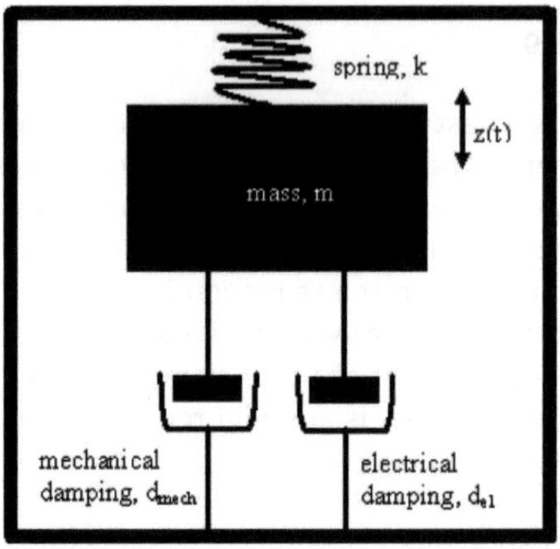

Fig. 6. General drawing of mechanical energy converter

Piezoelectric materials show an intrinsic electric polarization due to the crystal structure. An external electric field will lead to a mechanical deformation of the material (piezo effect) and vice versa a mechanical deformation will generate an electrical charge (inverse piezo effect). Thus, converting mechanical vibrations into deformations of the piezo material will generate an electrical voltage, which in turn can be used as power source. The energy density of a piezoelectric converter is strongly dependent on the coupling coefficient and the yield strength of the material. Roundy at al. have estimated the practical maximum energy density of piezoelectric converter to be approx. 17.5 mJ/cm^3 (Note: Assumed was a PZT-5H material with a factor of safety of 2). [7]

Electromagnetic conversion relies on the relative movement between an electrical coil and a magnetic field (permanent magnet). The change of the position due to vibrations leads to a current flow within the coil. The voltage on the coil is determined by Faraday's law can be used as power source. The energy density is strongly dependent on the magnetic field strength. Roundy at al. have estimated the practical maximum energy density of an electromagnetic converter to be approx. 4 mJ/cm^3 (Note: Assumed were a magnetic field of 0.1 T and the magnetic permeability of free space). [7]

Electrostatic conversion relies on the displacement between two electrical conductors separated by a dielectric material. Hence, the previously stored energy (electrical charges) changes with respect to the displacement. The voltage of a capacitor is given by the charge, the distance between both electrodes divided by the electrode area and the dielectric constant. If the voltage is held constant, the charge will increase with decreasing electrode distance, whereas if the charge is held constant, the voltage will increase with increasing electrode distance. In both cases the energy stored on the capacitor increases and can be extracted to power a device. Roundy at al. have estimated the practical maximum energy density of an electrostatic converter to be approx. 4 mJ/cm³ (Note: Assumed were an electric field of 30 MV/m (30 V/µm) and the dielectric constant of free space). [7]

A short summary of the different vibration harvesting concepts is shown in Tab. 4. As mentioned earlier, for different applications either one of the principles might be beneficial.

Type of energy conversion	*Advantages*	*Disadvantages*	*Comments*
Piezoelectric	*No voltage source needed* *Voltages > 1 V*	*Low coupling coefficient for thin film piezo* *Process integration?* *Reliability issues*	*High energy density for bulk material, but need for thin film processes*
Electromagnetic	*No voltage source needed*	*Voltages < 1 V* *MEMS integration?* *Permanent magnets and coils needed*	*Better performance at higher frequencies and limitations for micro scale*
Electrostatic	*MEMS integration* *Voltages > 1 V*	*Voltage source needed* *Parasitic capacitances* *Mechanical stops*	*Lower energy density, but fully MEMS compatible process*

Tab. 4. Summary of vibration harvester principles

Of importance for the harvester design are the frequency range and the amplitude of the excitations. In addition the limitations in the maximum volume/size due to package or weight issues are of interest for the design. In general the frequencies of the vibrations are in the range of 5 Hz–1 KHz depending on speed, tire size/model, suspension and road conditions. Most of the proposed vibration harvester concepts using inertia masses for storage of the kinetic energy. Thus, acceleration changes will be used to excite the harvester. From the published results the average power of a vibration harvester can be estimated to be in the range of several nW to a few mW depending on the design and volume of the harvester as well as on the amplitude and frequency of the vibrations.
For TPMS applications a power level several µW have to be achieved in order to start common power management circuits. In addition a temporary energy storage device such as a capacitor or secondary battery might be useful to provide additional energy for power consuming actions such as RF transmission or micro-controller processes.

5 Summary and Conclusion

Primary batteries will still be the major power sources for TMP systems within the next couple of years. However, technical advances concerning ultra-low power ASIC design and power saving algorithms will enable the use of either so-called hybrid systems or battery less energy harvesting devices. Different physical principles have been explored and could be used as power source for TMPS applications. The overall performance strongly depends on the specific application and design of the vibration harvester. Tire manufacturers and automotive system providers have started research projects within the field of battery less TPMS concepts.

References

[1] S.Hackl, "Trends bei Reifendruckkontrollsystemen (RDKS) Vom Komfort- zum Sicherheitsfeature", Sensoren im Automobil, 2006

[2] M.Fischer, "Tire Pressure Monitoring", Verlag Moderne Industrie, 2003

[3] National Highway Traffic Safety Administration, U.S. Department of Transportation, Federal Motor Vehicle Safety Standards, 49CRF571, NHTSA 2000-8572, 2004

[4] International Organisation for Standardization, Draft International Standard ISO/DIS 21750

[5] C.Raynaud, "SOI for low power" Minatec Crossroads, 2006

[6] T. Minami (Ed.), Solid State Ionics for Batteries, Springer-Tokio 2005, pp. 64-73

[7] S. Roundy, P.-K. Wright, J. M. Rabaey, Energy Scavenging for wireless sensor networks, Kluewer Academic Publisher, pp. 47-50

Thomas Lange, Dr. Markus Löhndorf
Infineon Technologies AG
Am Campeon 1-12
85579 Neubiberg
Germany
markus.loehndorf@infineon.com
thomas.lange@infineon.com

Terje Kvisterøy
Infineon Technologies SensoNor AS
Knudsrodvein 7
3192 Horten
Norway
terje.kvisteroy@sensonor.no

Keywords: tire pressure, battery-less, energy harvesting, power management

Thomas Lange, Dr Markus Löhndorf

Inertial Sensor Performance for Diverse Integration Strategies in Automotive Safety

E. Axten, J. Schier, Robert Bosch GmbH

Abstract

Vehicles today are often fitted with a number of inertial sensors which have different specifications and come from different suppliers but which measure the same inertial signals – angular rate and acceleration. Integration of these sensor elements into new or existing ECUs could reduce the number of sensors, bringing cost benefits without a deterioration in system performance. This paper will discuss this development, the requirements it places upon sensor elements and how these challenges can best be met. An assessment of current sensor elements and their suitability for use in different integration strategies completes the paper.

1 Inertial Sensor Elements

The next two points will introduce the SMG074 and SMG075 angular rate and SMB225 dual axis acceleration sensor elements. These sensors will be used, together with a leading competitor, as a benchmark for the standard of inertial sensor elements today. They will also be used to illustrate the benefits and limitations of the different system architectures introduced in Section 2.

Three different sensor elements were developed to cover the requirements of a large variety of systems: two angular rate sensor elements with measuring ranges of $\pm 187°/s$ (SMG074) and $\pm 244°/s$ (SMG075), and a dual axis acceleration sensor element with two measuring ranges: a low g, high accuracy measuring range of $\pm 4.9\ g$, and a high g measuring range of $\pm 38\ g$ but with lower accuracy. All of the sensor elements measure in-plane and offer excellent offset stability, low noise and a high level of electrical and mechanical robustness. The sensor elements are available in premold packages which are suitable for lead free and conventional leaded SMD soldering processes.

The BOSCH sensor elements offer the following advantages for system integration:

- Application circuit consists of very few external passive components (no capacitor required for output filter, for example)
- The digital SPI interface allows as many sensor elements to be easily connected to the vast majority of current microcontrollers. The only requirement is a synchronous serial interface in order to receive data from up to 8 sensing directions. The use of additional chip select lines increases the number of sensing directions up to the limits of the microcontroller.
- The sensor elements can be driven with 5 V or 3.3 V depending upon the requirements of the application, in order to ensure the integration in various ECU generations.

1.1 Angular Rate Sensor Elements

The angular rate sensor elements SMG074 and SMG075 use the CVG (Coriolis Vibratory Gyroscope) principle: values are acquired using capacitive effects and are directly converted into digital signals using a delta-sigma analog to digital converter. The MEMS structures have a double resonant working frequency of 15 kHz, which allows a high level of electrical and mechanical robustness to be attained.

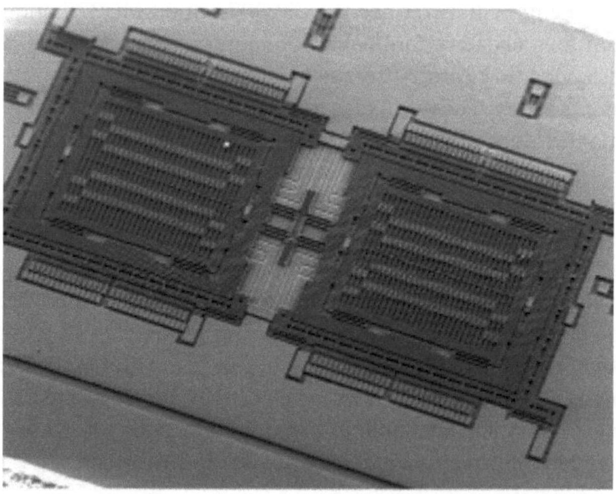

Fig. 1. Surface MEMS used in SMG07x (SEM-photo)

The signal processing ASIC is produced in a HVCMOS (High-Volt CMOS) process and has a 16 bit output resolution via SPI. A variety of monitoring signals and status bits are also transmitted via the SPI. Both sensor elements have periodic and commanded self test features. The ASIC and the micromechanical measuring element are packaged in a PM16 (premold 16) package as shown in Fig. 2 with a body size of 13.5 x 1 mm.

Fig. 2. Construction of the angular rate sensor element SMG074/5

The angular rate sensor elements are based upon the current production sensor (SMG070)[2], however due to the smaller ASIC they fit into a much smaller package – a space saving of about 40% – and thus the sensor elements are attractive for applications where space is a critical parameter.

Fig. 3. The angular rate sensor element SMG074

The SMG074 is mainly used for the measurement of vehicle yaw rate for ESP® systems. The SMG075 sets new goals for MEMS angular rates sensors with extended measuring ranges with respect to offset stability and noise. When mounted in different axes the SMG075 can be used in systems for improvements of roll and pitch stability and also in passive safety systems such as Roll Over Sensing (RoSe). In this way the SMG075 is the first step towards a sensor synergy between active and passive safety systems. It combines the increased measuring range (240°/s) required by passenger restraint systems and the higher accuracy (16 bit resolution and offset 1.5°/s) required by active safety systems. The importance of this application area is currently increasing due to the demand for SUV and van type vehicles with higher centre of gravity.

1.2 Acceleration Sensor Elements

The acceleration sensor element is a dual axis measuring element with dual axis ASIC. The MEMS structures are based upon the surface micromachining technology of current single axis production sensor (SMB220) [2], however the capacitive structures are realised twice orthogonal to one another in order to measure two axes simultaneously. The signal processing ASIC is also produced using the same HVCMOS process as the angular rate sensor ASIC. As the ASIC front end is also realised twice no multiplexing is required and therefore no performance constraints occur in the dynamic range of the signal. The yaw rate and acceleration sensor elements communicate using the same SPI protocol and are therefore fully compatible with one another. The two acceleration signals have an output resolution of 16 bit. In order to enable a multi level safety concept in the host ECU a large number of monitoring signals, status bits and self-test functions can be read out and commanded via the SPI interface. The measuring range of $\pm 4.9\ g$ for the high accuracy signal enables the use of the SMB225 for sensing in the z-axis. In this way a concept with redundant a_y and single a_x and a_z signals can be realised using just two identical acceleration sensor elements – an advantage for logistics.

Fig. 4. The dual axis acceleration sensor SMB225

The sensor element is constructed in the same way as SMG074 and SMG075, see Fig. 2, however like its predecessor SMB220 it is packaged in the smaller PM12 (premold 12) package with a body size of 11 x 11 mm. The SMB225 offers tried and tested technology and is twice as compact as the previous design.

2 Integration Strategies

Vehicle architectures are currently facing the challenge of incorporating an increasing number of functions and features, causing the number and complexity of electronic control units in the vehicle to explode. The networking between these ECU´s becomes more and more complex leading to a higher bus load, increased effort and cost in wiring and potential impacts on controllability. Advanced self diagnostics and enhanced safety integrity are an absolute necessity in these distributed logic systems, especially when the year on year increases in quality targets are taken into account. The increased number of ECU´s, sensors and their complexity also leads to an increase in material and manufacturing costs. New concepts are needed to meet the requirements of these future trends.

The following section reviews the integration strategies in use today, and those which show the most promise for future development. As more and more safety systems become commodities, the price is put under very high pressure and solutions requiring less expensive technologies and less space will be required. High end car manufacturers on the other hand are more motivated than ever to include more systems and to add sophisticated features to both safety relevant, and non relevant systems in order to distinguish themselves from their competitors.

2.1 Sensor Cluster of Inertial Sensor Elements for Vehicle Dynamic Systems

This is the most frequently found option in cars today. A sensor cluster consisting of different inertial sensor elements communicates with the ESP® ECU via CAN. Other ECUs can also use the signals – either by being directly connected to the CAN bus, or by receiving them from the ESP® ECU via a different connection. Due to its small size the sensor cluster can then be placed in a favourable position in order to measure the dynamics of the vehicle (usually near the middle), without excessive vibration. The system ECU can be placed wherever it needs to be – there are no restrictions placed upon it by the inertial sensors.

Fig. 5. DRS-MM3.x sensor cluster from BOSCH

2.2 Integration of the ESP® Inertial Sensors into the ESP® ECU

The inertial sensor elements can also be integrated into the ESP® system ECU. This offers the advantage of having no extra connection (wiring harness) connecting the sensors and the ECU. In addition no extra sensor housing is required, and so no suitable sensor mounting position must be found. Costs may also be saved by not requiring the additional microcontroller and CAN circuitry found in the sensor cluster. The main challenge associated with this concept is the comparatively harsh environment as the ESP® ECU is located in the engine compartment.

Fig. 6. Integration of inertial sensor elements into the ESP® ECU

2.3 Integration of the ESP® and Airbag System Inertial Sensors into the Airbag ECU

As the airbag ECU already contains inertial sensor elements, its mounting position and housing concept have been developed with these requirements in mind and there are opportunities for sensor synergies. The addition of inertial sensor elements for vehicle dynamic control saves on space in the vehicle as a whole as the sensor cluster must not be placed, however may lead to space problems in the Airbag ECU itself. As in the previous example the separate sensor housing is no longer required.

Fig. 7. Integration of inertial sensor elements into airbag ECU

2.4 Integration of all Inertial Sensor Elements at Vehicle Level in a Domain ECU, including Active and Passive Safety Systems

This concept involves the integration of all inertial sensors in the vehicle into one central ECU – this means that the signals for a variety of system ECU´s (ESP, automatic cruise control, roll stability control, airbag, etc.) can be housed together. In this way the dynamic behaviour of the vehicle can be completely measured and processed using model based algorithms. The required signals can then be supplied to all active and passive safety chassis and restraint systems. This has the advantage, that the requirements of the inertial sensor elements can be taken into account when choosing a mounting position, and there is no competition between two ECU´s or sensor clusters that both require a mounting position suitable for Vehicle Dynamics Management (VDM). Synergy effects between the sensor elements, and the signals required by active and passive safety systems can be used to the full. If the required systems use all inertial sensing axes, the ECU can be mounted in an arbitrary position. The misalignment with respect to the vehicle axes can then be calculated and automatically corrected for.

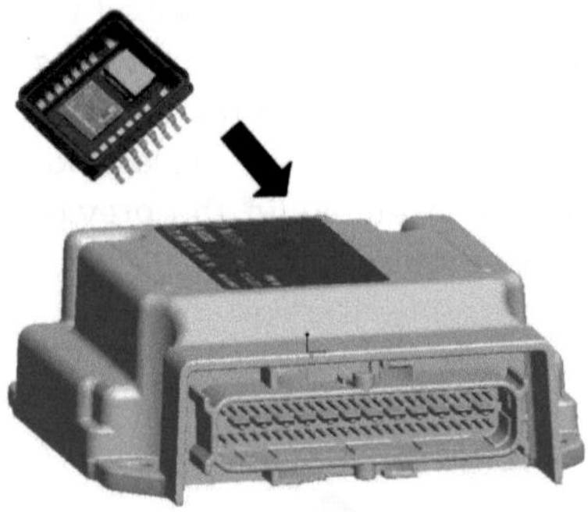

Fig. 8. Integration of inertial sensor elements into a domain ECU

The main challenge in using this integration method to house airbag sensors is that the energy reserve requirements for restraint systems must be met. This means that the airbag sensors in the domain ECU require a longer buffer time. Concepts involving intelligent energy management and interfacing to the airbag logic are currently in development.

With the introduction of FlexRay® bus systems in new vehicle generations it is possible to achieve higher and higher data rates and more networking of different systems. This increases the attractiveness of a centralised domain ECU to host inertial sensors.

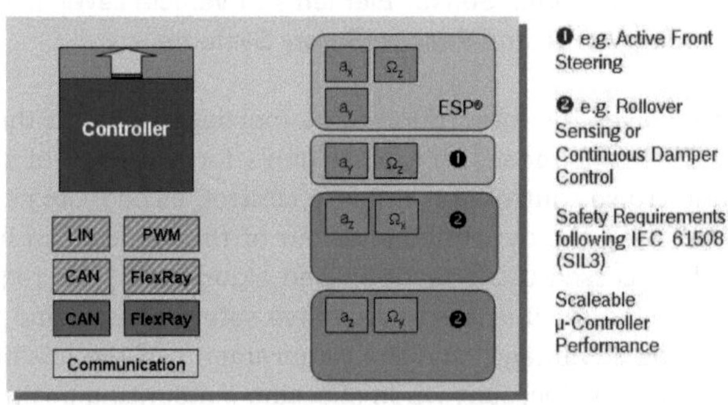

Fig. 9. Functional block diagram of a domain ECU

3 Requirements on Inertial Signals

Just to meet the sometimes contradictory demands of active and passive safety systems is a challenge in itself. If one considers the different purposes to which the sensors are put. In order to be suitable for use in a wide range of integration concepts and ECUs, inertial sensor elements must be as flexible as possible. This means flexible interfaces – different supply voltages, digital interfaces, programmable interfaces, robust packaging, electrical robustness and so on. What is however also required is that the sensor elements meet the specification requirements of the most demanding system they will supply signals to. This may sound banal, however it is important to realise, that when we talk about integration as a method of saving money, or space, a logical consequence is that the demand for high performance sensor elements will rise. It is possible to use an ESP gyro for navigation systems or artificial horizon displays within the vehicle – it is not possible to use the consumer products usually associated with these applications for active and passive safety systems.

This section will deal with demands placed upon the inertial sensor elements discussed in section 1, by the integration strategies discussed in section 2.

3.1 Comparison of Requirements for the Different Integration Concepts

Active and passive automotive safety systems both use inertial sensor elements for monitoring of vehicle dynamics, however the exact requirements of the different systems vary greatly according to how these signals are to be used. An airbag acceleration sensor for example, has its primary purpose in detecting a crash. To do this it must be able to detect high accelerations, or sudden large changes in acceleration without being damaged by them and must not pose a considerable load on the airbag energy reserve after the battery has been disconnected. Requirements for signal offset and zero point error are often achieved by an automatic offset correction at the output stage. Sensitivity error tolerances are also generous in comparison to requirements of active systems.

The requirements for active systems however show a different focus: the signal range is smaller and it is the task of the sensor to reproduce the inertial signal faithfully without distortion. Thus the demands for accuracy and stability over lifetime are considerably higher on these sensors, and it sets the standard concerning performance for any synergy products.

It is possible to group the requirements on the inertial sensor elements into two main areas – environmental considerations and performance specifications. For synergy projects to work successfully the specification for the sensor elements is dictated by the tougher specification. This leads to higher overall specification levels for synergy projects, and increases the attractiveness of high specification sensor elements even if they are more expensive than those traditionally used for other systems. The following table summarises the different environmental constraints imposed by the integration concepts. Where no special requirement exists a '-' is shown.

	Sensor cluster	*ESP® ECU integration*	*Airbag integration, Roll over functions*	*Domain Control Unit, High end*
Supply voltage	*Can be chosen according to Sensor elements' req.*	*Sensor Vs dictated by ECU*	*Sensor Vs dictated by ECU*	*Sensor Vs dictated by ECU*
Current	*Voltage regulator selected according to current req.*	-	*Low current due to required energy reserve*	*Low current due to required energy reserve*
Temperature range	*Passenger compartment*	*Engine compartment*	*Passenger compartment*	*Passenger compartment*
g-sensitivity angular rate sensor	*Passenger compartment, small box volume enables flexible mounting location*	*Engine compartment, fixed mounting position due to ECU/HU*	*Passenger compartment, mounting position optimised for Airbag inertial sensors*	*Passenger compartment, can be placed in former Airbag mounting position*
Sensor electronic interface	*Digital or Analog, due to limited sensing axes*	*Digital or Analog, due to limited sensing axes*	*Digital, Bus due to multiple sensing axes*	*Digital, Bus due to multiple sensing axes*
Space constraints	*Housing can be constructed according to sensor size*	*Small size req. to fit in existing housing*	*Small size may be req. to fit in existing housing*	*Small size may be req.*

Tab. 1. Environmental requirements of different integration concepts on inertial sensor elements

The sensor cluster shows the current state of the art technology and has few special requirements on the sensor elements, as the cluster is specially designed to meet the needs of the elements. The ECU´s are generally not designed with the sensors in focus, and therefore the sensor elements must be more robust and more flexible. The requirements of the synergy systems (airbag integration and domain control unit) are noticeably different to those posed by the ESP® ECU.

	Sensor cluster	*ESP® ECU integration*	*Airbag integration, Roll over functions*	*Domain Control Unit, High end*
Measuring range	*75°/s, 1.2 g (ESP) 120°/s, 2.5 g (advanced active systems)*	*75°/s, 1.2 g*	*Dual/extended range*	*Extended range*
Sensitivity error	*5%*	*Reduced requirements (correction of ECU misalignment in vehicle provided)*	*5%*	*Increased requirements*
Offset (g)	*100 mg*	*Reduced req. (correction of ECU misalignment in vehicle provided)*	*100 mg*	*Increased requirements**
Offset (°/s)	*3.5°/s*	*3.5°/s*	*Increased requirements**	*Increased requirements**
Offset stability (g)	*70 mg*	*70 mg*	*70 mg*	*Increased requirements*
Offset stability (°/s)	*2°/s*	*2°/s*	*Increased requirements*	*Increased requirements*
Offset drift due to temp (g)	*30 mg*	*Increased requirements (larger T-range)*	*30 mg*	*Increased requirements*
Offset drift due to temp (°/s)	*1.5°/s*	*Increased requirements (larger T-range)*	*Increased requirements*	*Increased requirements*
Rate of change of offset	*1.5°/s/min, 30 mg/min*	*1.5°/s/min, 30 mg/min*	*Increased requirements*	*Increased requirements*
cross axis sensitivity	*5 %*	*Reduced req. (correction of ECU misalignment in vehicle provided)*	*Increased requirements**	*Increased requirements**
Output signal dynamic range	*50 Hz*	*Low*	*High*	*High*

Tab. 2. Performance Requirements of different integration concepts on inertial sensor elements

Increased requirements: * for systems with 6D performance can be improved using software to calculate and correct misalignment in all axes.

Signal performance requirements for synergy systems are highest – the challenges for ESP® ECU integration are based upon the harsh environment.

3.2 Strengths and Weaknesses of the Inertial Sensor Elements, based on use in the Various Concepts

Tab. 3 evaluates the sensor elements according to their suitability for use in the various integration concepts.

Sensor element	*Increased system requirements met?*					
	ESP ECU		*Airbag ECU*		*Domain Control unit*	
	environmental	*performance*	*environmental*	*performance*	*environmental*	*performance*
SMG074	*No**	*Yes*	*Yes*	*Yes*	*Yes*	*Yes*
SMG075	*No**	*Yes*	*Yes*	*Yes*	*Yes*	*Yes*
SMB225	*No**	*Yes*	*Yes*	*In Part***	*Yes*	*Yes*
Competitor accleration	*No*	*n.a.*	*In Part (no flexible supply)*	*No (limited measuring range)*	*In Part (no flexible supply)*	*No (Temp. Offset Drift)*
Competitor angular rate	*In Part*	*Yes*	*Yes*	*Yes*	*Yes*	*In Part*

Tab. 3. Suitability of sensor elements for system integration

* The SMB and SMG in the form discussed in the paper are not suitable for ESP® ECU integration as the packaging is only qualified to 120°C. A modification for harsh environments is in development. ** SMB225 covers all low-g specification requirements

The suitability of the SMG and SMB for sensor synergy concepts as shown in the table shows a high degree of future proofing.

4 Summary and Conclusion

As shown in this paper the increasing diversity in inertial sensor concepts, and their corresponding requirements lead to an increase in the specification requirements of inertial sensor elements. In order for the higher levels of integration to function successfully a certain degree of sensor synergy must be obtained- and in this case the application with the highest performance will always dictate the sensor performance requirements. SMG and SMB inertial sensor elements from BOSCH provide an excellent solution for increasingly complex integration strategies, and support sensor synergy projects.

5 Acknowledgements

The authors would like to thank all their colleagues at Automotive Equipment, Division Automotive Electronics and Division Chassis Systems for the design, development and testing of the complete sensor system, the design and layout of the sensor measuring elements and sensor readout electronic modules, and for the mechanical construction as well as for the design and implementation of hard- and software.

References

[1] J. Schier, R. Willig, K. Miekley, „Mikromechanische Sensoren für Fahrdynamik-Regelsystem", Automobiltechnische Zeitschrift 11/2005, 107. Jahrgang, November 2005, S. 992

[2] J. Schier, R. Willig, "New Inertial Sensor Cluster for Vehicle Dynamics Systems" AMAA 2005

E. Axten, J.Schier
Robert Bosch GmbH
Robert Bosch Allee 1
D-74232 Abstatt
emma.axten@de.bosch.com
johannes.schier@de.bosch.com

Keywords: inertial sensor cluster, yaw rate sensor, angular velocity sensor, acceleration sensor, silicon surface micromachining, modular concept, signal accuracies and characteristics, cross-system application, integration, ESP, HHC, ACC, 4w, ROM, RoSe, EAS, steer-by-wire

Far Infrared Low-Cost Uncooled Bolometer for Automotive Use

T. Kvisterøy, H. Jakobsen, Infineon Technologies SensoNor AS
C. Vieider, St. Wissmar, P. Ericsson, U. Halldin, Acreo AB
F. Niklaus, Fr. Forsberg, G. Stemme, Royal Institute of Technology
J. E. Källhammer, H. Pettersson, D. Eriksson, Autoliv Research
J. Franks, J. VanNylen, H. Vercammen, Umicore Electro-Optic Materials
A. VanHulsel, Vito

Abstract

A proposed EU regulation requires the automotive industry to develop technologies that will substantially decrease the risk for vulnerable road users such as pedestrians when hit by a vehicle. Automatic brake assist systems, activated by a suitable sensor, will reduce the speed of the vehicle before the impact. Far InfraRed (FIR) detectors are ideal candidates for such sensing systems. In order to enable high volume serial installation, the main development must be focused on cost reduction. Optimizing all aspects of the system, including sensor size, production yield and 3D wafer level vacuum packaging will lower today's "high end" FIR product costs by an order of magnitude. A low-cost FIR infrared bolometer is developed in the Eureka labeled PIMS (Pedestrian Injury Mitigation System) project. In the paper the background and the actual design of the first demonstrator to be finished in 2008 are described in detail.

1 Introduction

Pedestrian fatalities are around 24% of all traffic fatalities worldwide [1] and about 15% of the traffic fatalities in Europe [2]. The European Commission is aiming at decreasing fatalities in traffic and specifically improving safety of pedestrian and other vulnerable road users (VRU) through an EU directive that requires the automotive industry to develop technologies that will substantially decrease the risk for a pedestrian or cyclist to be killed or seriously injured when hit by a car.

A new draft proposal for the regulation was issued in 2005, which opens up the possibility of fulfilling the regulation by means of collision mitigation measures such as pre-impact braking. As the proposed regulation also stipulate that so called brake assist systems will be mandatory, all new vehicles will have the necessary actuator for autonomous pre-impact braking if augmented with a suitable sensor. Brake assist is a system that reduces the braking distances by applying full braking power in situations identified as a panic stop.

The current functionality of brake assist systems has been estimated to greatly reduce the pedestrian fatality rate according to a study by Transport Research Laboratory (TRL) under contract by the European Commission [3]. The TRL study showed that when comparing the current phase 2 of the Pedestrian Protection Directive with the effects of a regulation with mandatory brake assist systems, as proposed by the European Automobile Manufacturers Association (ACEA), additional 16% pedestrian fatalities would be saved. A brake assist system is, however, dependent on a rapid reaction by the driver.

The availability of brake assist systems in all cars enable also autonomously applied braking when complemented with a suitable sensor. Such a system is expected to comply with the collision mitigation measures alternative proposed in the above mentioned EU proposal. An automatic brake assist system can reduce the speed of the car before an impending pedestrian impact independent of any driver interaction.

A system that automatically brakes the vehicle when a risk of impact is detected, will significantly reduce the fatalities and level of injuries. For a car travelling at 50 km/h it takes 0,72 s to move 10 m. Activating the brakes with a mean level of 0.6 g when the car is 10 m before the impact with the pedestrian reduces the impact velocity to 31 km/h. Accident data from the International Harmonization Research Activity (IHRA) [3], indicate that this would reduce the risk of fatality from about 35% to about 10%, which can be seen in Fig. 1. As the driver either does not react at all, or does not apply the brakes with sufficient rate to activate the brake assist system in 53% of the pedestrian accidents according to a study of the University of Dresden [4], it implies that the fatality reduction potential with an autonomously activated brake assist system would be up to double the estimated 16% of current brake assist systems.

There are at least two advantages of pre-impact braking: The impact energy will be reduced significantly and the common secondary impact when the pedestrian hit the street will also be reduced. Pre-impact braking is the only protection method for pedestrians to reduce the second impact [5].

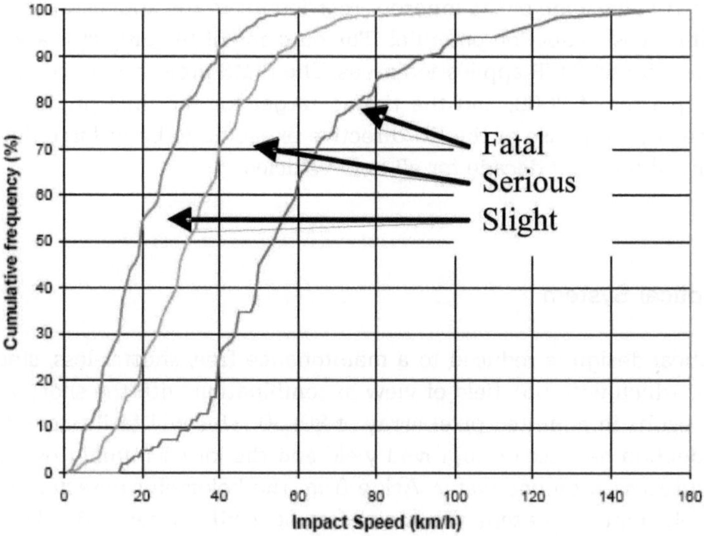

Fig. 1. Pedestrian fatality and injury risks as a function of impact speed [3]

Necessary for a realistic realization of an autonomous pedestrian protection brake assist system is of course the availability of a robust and cost effective sensor system that can be afforded as serially installed equipment. Far infrared (FIR) thermal cameras are particularly well suited to detect human beings and animals, who present high, signal to noise ratio versus ambient due to their higher temperature. These cameras should make the picture less difficult to interpret for an automatic vision algorithm, thus allowing for a less complex and therefore lower cost image processing hardware. An appreciable advantage of FIR systems is that they do not need an external illumination source and are also immune to the blinding effects by other illumination sources such as the headlights of oncoming cars.

However, the high cost of today's systems has restricted the use of FIR [6]. The proposed revised EU directive may generate the required high-volume application that will justify the cost reduction effort. This type of system is also intended to detect objects at a close range, which should permit a lower resolution. The distance between the camera and the object will typically be up to 15 m for processed pedestrians and other VRUs. The project reported here, Pedestrian Injury Mitigation System (PIMS) funded under the Eurimus/ EUREKA program [7], is aiming to design a FIR sensor optimized for the low cost, high-volume, short range requirements that can take advantage of the recent progress in Micro Electro-Mechanical Systems (MEMS) sensor development [8-10] and some of the inherent restrictions in the application, providing a ten times cost reduction potential. The outcome of the project is also expected to benefit other FIR application areas. The PIMS project will end during the second quarter of 2008, and the timing target for the final product aims to fulfil the second phase of the EU directive expected to be in force during the first part of the next decade for all new vehicles.

2 Optical System

The optical design is reduced to a maintenance free, shutter-less, single lens camera, which with 55° field of view in combination with the short detection range, results in a limited pixel array of 80x30. This will facilitate a dramatic cost reduction because of improved yield and the larger number of chips that can be processed on one wafer. Aside from the bolometer detector array, the optical elements represent the main cost of a FIR system. The key optical parameters that need to be extracted are the focal length of the optic and the *f*/number of the system. The focal length is defined by the field of view required and the size of the detector. With simple trigonometry the focal length then can be determined; (see Fig. 2).

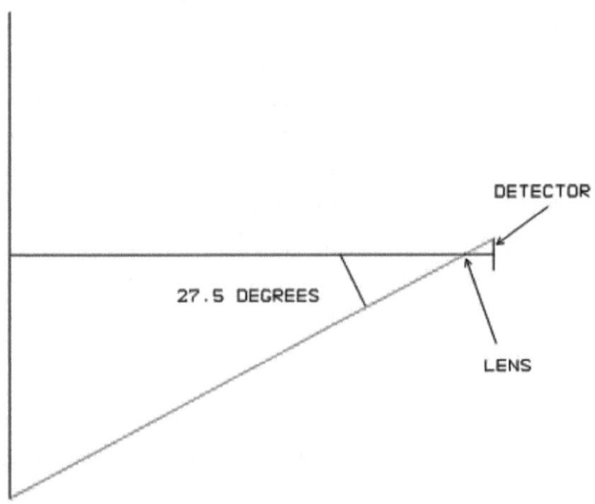

Fig. 2. Optical system layout.

The *f*/number is more complicated since there is a trade off between optical cost and complexity and the detector cost and complexity. Simply stated a fast optical system is more costly, and a more sensitive detector is more costly. Since neither parameter is fixed in any absolute sense, it becomes necessary to find the best compromise. A step change in cost arrives when it becomes necessary to use two lens elements to meet the optical resolution requirements. Trial solutions have therefore been made to determine the maximum *f*/number which can be achieved with a single element. Beyond the cost of optical manufacture, the advantage of a single element solution is in reduced metalwork costs and assembly times. In order to achieve a sufficient level of energy collection with a single element lens it would be necessary to make both the surfaces aspheric.

Both germanium and the chalcogenide glass GASIR®3 from Umicore were investigated as potential lens materials. Germanium is a high-performance material for FIR lenses, with a high mechanical resistance and a refractive index of 4.0. It is, however, expensive and requires a costly single-point diamond turning process, to machine it into an aspherical surface. GASIR®3 is a new type of infrared transmitting glass with a refractive index of 2.5. It is an attractive candidate for high-volume infrared lenses as it can be moulded into a finished lens. In addition, the highly variable germanium market price affects GASIR®3 to a lesser extent because it contains less then one quarter germanium. GASIR®3, unlike germanium, also possesses a refractive index that does not vary much with temperature, allowing for a simple opto-mechanical design that does not require temperature compensation. One drawback of chalcogenide glasses like GASIR®3 is their higher chromatic dispersion as compared with germanium. To overcome such dispersion, optics with diffractive or binary compensation must be fabricated. Fortunately, these specially shaped surfaces can be incorporated during the GASIR®3 moulding process at no additional cost. It was found that it would be possible to have a single element solution in both materials; (see Fig. 3, 4). When performance at the high and low temperature extremes is taken into account, there is no clear preference for either material in terms of the optical performance.

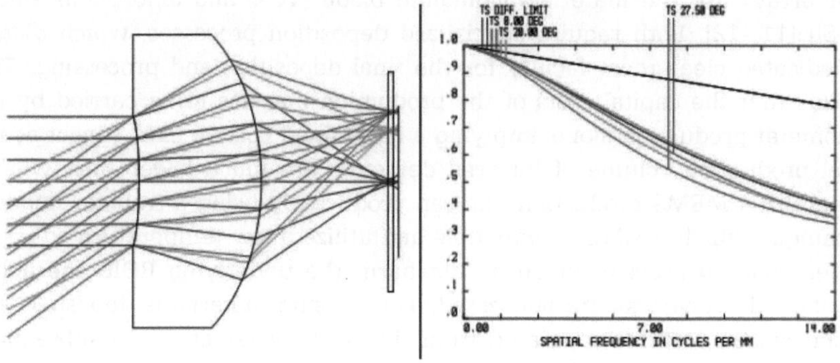

Fig. 3. GASIR®3 solution.

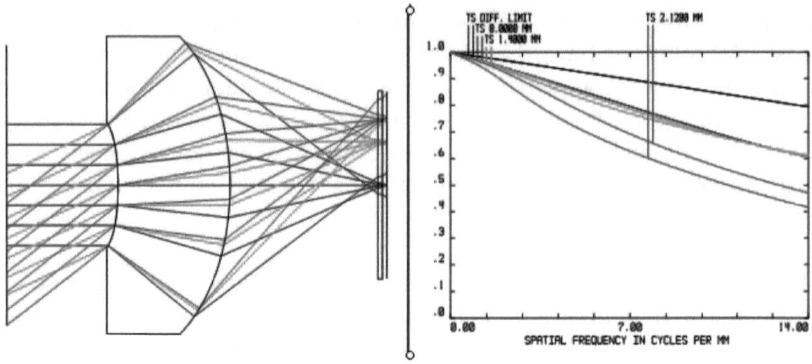

Fig. 4. Germanium solution.

3 FIR Detector

To be able to reach the cost target a significant cost reduction compared with the presently used "high-end" bolometer array fabrication process is needed. Conventional infrared bolometer arrays are monolithically integrated on top of the Read Out Integrated Circuit (ROIC) wafers. The main drawback of such design is that only CMOS compatible fabrication and material deposition processes can be used for the infrared bolometers. Temperatures above about 450 °C will harm standard CMOS and thus, cannot be used for monolithically integrated infrared detectors. Current materials for use below 450 °C suffer from large variation in electrical parameters of the bolometers and lower production yield. In addition, optimization of the high temperature coefficient of resistance and low $1/f$ noise parameters are difficult. Current infrared bolometer arrays use the materials vanadium oxide (VOx) and amorphous silicon (α–Si) [11, 12]. Both require specialized deposition processes, which dictate a dedicated clean-room facility for the final deposition and processing. This means that the capital costs of the production line has to be carried by the bolometer production alone, implying a large fixed cost on each sensor as the total production volume of infrared detectors are much lower than what a generalized MEMS production line can produce. By using a transfer bonding technique which is CMOS compatible and utilize a low temperature adhesive wafer bonding process which do not harm the underlying ROIC the above mentioned drawbacks are addressed. The sensing material is deposited and optimized on a wafer separated from the ROIC wafer and in a subsequent step the detector material is transferred from the detector substrate wafer to the ROIC wafer. The detector wafer will then be thinned down leaving only the detector material. The final bolometer design and function is identical to monolithically fabricated bolometers but using a crystalline sensing material. This approach allows us to independently optimize the processing of the ROIC and of the bolometer material and to manufacture them in existing MEMS foundries in very high volumes.

3.1 Quantum well Material

The detector array is based on a new type of high performance thermistor material. Very thin Si/SiGe single crystal multi-layers are grown epitaxially. These quantum well structures confines particles to a planar region, forcing electrons- which were originally moving freely in three dimensions- to occupy only two dimensions. This in turn creates discrete energy levels known as "energy sub-bands". This confinement comes into place once the thickness of the quantum well is in the same range as the "de Broglie" wavelength. For the PIMS application, the quantum wells functions as the thermistor material, changing its resistance as the temperature varies. The more the resistance changes for each degree of temperature change, the better its sensitivity. This is quantified by the temperature coefficient of resistivity, TCR. The temperature coefficient is defined as:

$$\beta = \frac{1}{R}\frac{\partial R}{\partial T} \quad (1)$$

where R is the resistance and T is the temperature. For a semi conducting material, $1/R$ is proportional to the amount of free carriers, p_{exc} (as in this case of p-doping) [13], hence the temperature coefficient can be calculated and derived to the quantum well's barrier- and fermi energy level by:

$$\beta = -\frac{1}{k_B T^2}\left(\frac{3}{2}k_B T + E_f - V\right) \quad (2)$$

where k_B = Boltzmann's constant, V= barrier energy, E_f= fermi level and T = temperature.

In Fig. 5 a schematic diagram of the valence band (p-type) structure of a single quantum well is shown. By varying the germanium content the barrier height, V, can be changed. The fermi level can be controlled by the quantum well width and the doping level of the well. In this way the material can be tailored to maximize the needed characteristics. Theoretically a TCR value of 4.2%/K can be achieved with this particular material system. Experimentally so far a value of 2.0%/K has been obtained [13]. Simultaneously, the use of high quality crystalline material provides very low 1/*f*-noise characteristic and uniform material properties over the full automotive temperature range. The inherent Noise Equivalent Temperature Difference (NETD) is expected to be better than for VOx and amorphous silicon bolometers. From a manufacturing point of view the well defined and uniform material properties obtained by mono crystalline deposition results in high yields compared to conventional uncooled detectors [14].

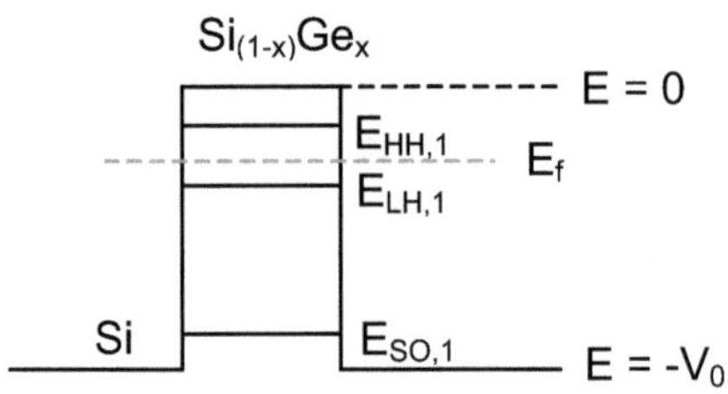

Fig. 5. Valence band structure of SiGe/Si quantum wells.

2.2 Bolometer Manufacturing Process

To integrate the high-performance SiGe multilayer structure on top of the CMOS based Read- Out Integrated Circuit (ROIC) a three-dimensional (3D) bolometer integration process is used [8-10]. In this 3D bolometer integration process the epitaxially grown SiGe multilayer structure is transferred from the original handle wafer to the ROIC wafer. The transfer from the handle wafer to the ROIC wafer is done using low-temperature adhesive wafer bonding in combination with sacrificial removing of the handle wafer as described in Fig. 6. This 3D bolometer integration technique is compatible with standard CMOS wafers and MEMS foundry processes and only makes use of standard semiconductor materials.

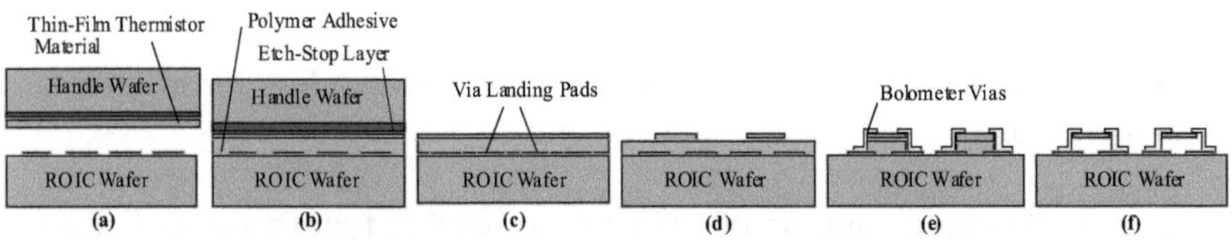

Fig. 6. 3D integration for uncooled infrared bolometer arrays: (a) fabrication of ROIC wafer and separate handle wafer with thermistor material (b) adhesive wafer bonding (c) thinning of handle wafer (d) bolometer definition (e) via formation (f) sacrificial etching of adhesive.

Fig. 7a shows a 3D drawing of one bolometer pixel with 35 µm long and 1.5 µm wide bolometer legs that provide low thermal conductance between the bolometer and its surroundings. The bolometer pixel pitch for the PIMS demonstrator array is 40 µm x 40 µm and contains 16 x 16 bolometer pixels. Fig. 7b shows a cross-sectional image of the bolometer structure with the SiGe multilayer material, the top and bottom contact electrodes and the bolometer leg structure. The used bolometer design has an optical cavity structure that is part of the bolometer membrane for improved absorption of infrared radiation. For the demonstrator array, a ROIC design is used that contains amplifiers and on-chip A/D converters to facilitate interfacing from the ROIC to the camera system. The demonstrator ROIC is based on a commercially available 0.8 µm BiCMOS process [15].

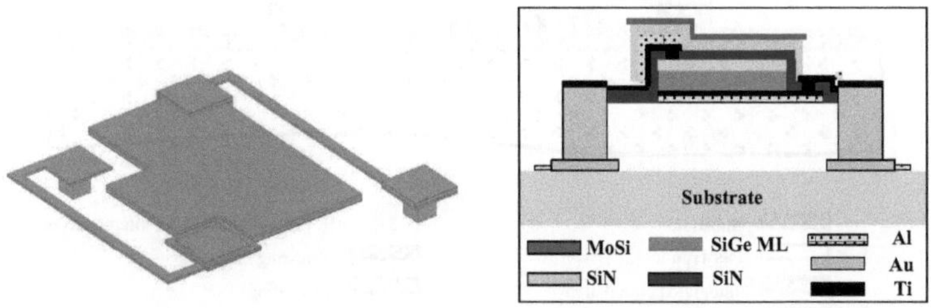

(a) (b)

Fig. 7. 3D drawing (a) of a bolometer pixel and cross-sectional image (b) of the bolometer structure with the top and bottom electrode contacting the SiGe multilayer structure.

4 Wafer Scale Packaging

Reliability and cost of micro systems is heavily related to the encapsulation techniques chosen. In the PIMS project we develop a wafer scale package to form lids with IR-window over the bolometer pixel array. This technology allows the pixels to be packaged in a hermetic cavity with the electrical contacts outside the cavity at low cost. As a result, this also allows the bolometer chip to be integrated directly to the camera housing, hereby reducing the system cost even more. The inherently good detector performance developed in the PIMS project allows us to us a lower vacuum without compromising overall performance too much. This enables good long term stability and makes the use of getter materials and custom processes obsolete.

The specified requirement, set to be <10mbar, is in line with many other MEMS devices on the market. In contrast conventional bolometers operate with vacuum levels $<10^{-3}$mbar [16, 17].

Wafer scale packaging technology has already proven its ability to secure vacuum and high reliability in other demanding applications like automotive Tire Pressure Monitoring (TPM) and Gyro's for the last 10 years [18]. Fig. 8 shows the principles of the wafer scale packaging technology that is being developed.

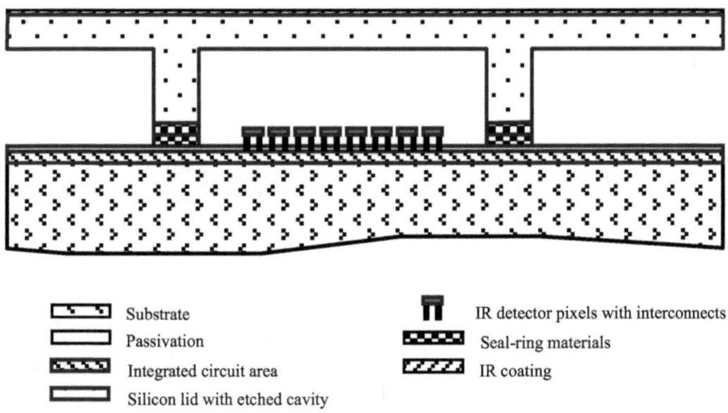

Fig. 8. The principle of how the cap-wafer is attached to the ROIC wafer with the pixels inside a separate cavity sealed in vacuum. The areas on the cap-wafer outside the cavity, is thereafter removed with a wafer dicing saw

The cap-wafer material (window) is made in standard single-crystal silicon wafers. Thin window areas are made by deep reactive-ion etching. Standard antireflection coating is deposited to obtain windows without any significant degradation of the final effective NETD. A seal ring is added on the frame areas of the ROIC wafer; (see Fig. 9). The cap-wafer is then aligned and attached to the ROIC wafer with the pixels and sealing take place in vacuum. An overview of commonly used wafer bonding techniques is given in [10].

After sealing, the areas on the cap-wafer around the pixel areas are removed by standard wafer dicing saw in order to release the wire bonding pads. No special planarization processes are needed on the ROIC wafer for the wafer level sealing process. The process sequence for the wafer scale packaging allows building of additional thin-film getter structures into each device. Hereby vacuum pressure $<10^{-3}$ mbar can be obtained and open up for applications requiring better NETD. Development of specific methods to measure the quality of the vacuum both on wafer level and in the final devices are investigated in the PIMS project. Establishing manufacturing schemes by using accelerated testing for checking the actual gas pressure and bonding consistency will hereby be made possible at low cost.

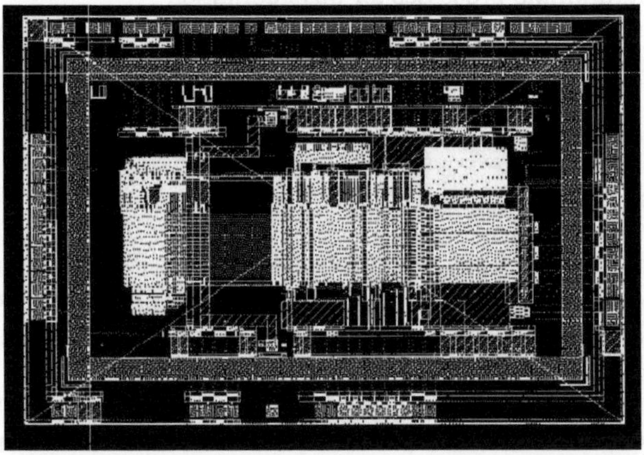

Fig. 9. Demonstrator ROIC pixel array with seal ring

5 Summary and Conclusion

A proposed EU regulation requires the automotive industry to develop technologies that will substantially decrease the risk for vulnerable road users such as pedestrians when hit by a vehicle. Automatic brake assist systems, activated by a suitable sensor, will reduce the speed of the vehicle before the impact, independent of any driver interaction. FIR detectors are ideal candidates for such sensing systems. In order to enable high volume serial installation, the main development must be focused on cost reduction. Optimizing all aspects of the system, including sensor size, production yield and 3D wafer level vacuum packaging will lower today's "high end" FIR product costs by an order of magnitude. Major cost reduction measures addressed in the PIMS project are:

- reduced features optical design with 15 meter range and 55° field of view
- maintenance free, shutter-less, single lens camera
- diffractive or binary compensated GASIR®3 moulded lenses
- less complex image processing hardware (due to use of FIR)
- limited pixel array of 80x30
- standard epitaxially grown SiGe multilayer quantum well sensing material on separate wafer

- standard CMOS (or BiCMOS) ROIC wafer
- adhesive low temperature wafer bonding to transfer the sensing material to the ROIC wafer
- reduced vacuum requirement to <10 mbar
- standard MEMS processes for pixel definition
- wafer scale packaging similar to present high volume MEMS automotive manufacture

References

[1] SAVE-U EU Project Deliverable 1-A, "Vulnerable Road User Scenario Analysis", http://www.save-u.org/download/PDF/D1A_V4.pdf, downloaded 15 Feb. 2006.

[2] D. M. Gavrila, "Sensor-based Pedestrian Protection", IEEE Intelligent Systems, Vol. 16 (6), pp.77-81, 2001.

[3] G. J. L. Lawrence, B. J. Hardy, J. A. Carroll, W. M. S. Donaldson, C. Visvikis, and D. A. Peel, "A study on the feasibility of measures relating to the protection of pedestrians and other vulnerable road users – Final report", http://europa.eu.int/comm/enterprise/automotive/pagesbackground/pedestrianprotection/pedestrian_protection_study.pdf, EU Contract no FIF. 20030937, downloaded 15 Feb 2006.

[4] Hannawald L., Krauer F., "Equal effectiveness study on pedestrian protection", Dresden, Germany: Technische Universität Dresden, http://europa.eu.int/comm/enterprise/automotive/pagesbackground/pedestrianprotection/summary_on_effectiveness.pdf, downloaded 15 Feb 2006.

[5] Meinecke M.M, Obojski M.A., Gavrila, D., Marc E., Morris R., Töns M., Letellier L. "Deliverable D6: Strategies in Terms of Vulnerable Road User Protection" http://www.save-u.org/download/PDF/D6_V3.0.pdf, downloaded 15 Feb 2006.

[6] J.J. Yon, L. Biancardini, J.L. Tissot, L. Letellier, "Infrared microbolometer sensors and their application in automotive safety", Proc. AMAA 2003, pp.1-15, Berlin, Germany, http://www.save-u.org/download/PDF/AMAA_200305.pdf, 2003.

[7] The Eurimus/EUREKA web site, http://www.eurimus.com/index.php?flashOK=10

[8] F. Niklaus, J. Pejnefors, M. Dainese, M. Haggblad, P.-E. Hellström, U.J. Wallgren, G. Stemme, "Characterization of transfer-bonded silicon bolometer arrays", Proc. SPIE, Vol. 5406, pp. 521-530, Orlando, USA, 2004.

[9] F. Niklaus, E. Kälvesten, G. Stemme, "Wafer-level membrane transfer bonding of polycrystalline silicon bolometers for use in infrared focal plane arrays", Journal of Micromechanics and Microengineering, Vol. 11, pp. 509-513, 2001.

[10] F. Niklaus, G. Stemme, J.-Q. Lu, R.J. Gutmann, "Adhesive wafer bonding", Journal of Applied Physics: Applied Physics Reviews, Vol. 99, No. 3, pp.031101/1-031101/28, 2006.

[11] M. Kohin, N. Buttler, "Performance limits of uncooled VOx microbolometer focal plane arrays", Proc. SPIE 2004, Vol. 5406, pp.447-453, Orlando, USA.

[12] E. Mottin, A. Bain, J.L. Martin, J.L. Ouvrier-Buffet, S. Bisotto, J.J. Yon, J.L. Tissot, "Uncooled amorphous silicon technology enhancement for 25 µm pixel pitch achievement", Proc. SPIE 2003, Vol. 4820, pp. 200-207, Seattle, USA.
[13] S.G.E. Wissmar, L. Höglund, J.Y. Andersson, C. Vieider, S. Savage, P. Ericsson; "High signal-to-noise ratio quantum well bolometer materials"; Proc. SPIE Vol. 6401, 64010N, Optical Materials in Defence Systems Technology III.
[14] P.W. Kruse, "Can the 300 K radiating background noise limit be attained by uncooled thermal imagers?", Proc. SPIE, Vol. 5406, pp.437-446, Orlando, USA, 2004.
[15] C. Jansson, U. Ringh, K. Liddiard, N. Robinson, "FOA/DSTO uncooled IRFPA development", Proc. SPIE 1999, Vol. 3698, pp. 264-275, Orlando, USA.
[16] J.J. Yon, L. Biancardini, J.L. Tissot, L. Letellier, "Infrared microbolometer sensors and their application in automotive safety", Proc. AMAA 2003, pp.1-15, Berlin, Germany.
[17] XenIC preliminary datasheet: "8 - 10µm Micromachined Bolometer", March 2003, Belgium.
[18] Roy Grelland, Christian Burrer, Matthias Halsband, Ronny Weum, "Direct measuring Tire Pressure Monitoring Systems", Sensors Expo, Chicago 2004.

Terje Kvisterøy, Henrik Jakobsen
Infineon Technologies SensoNor AS
Knudsroedveien 7, P.O.Box 196
3192 Horten
Norway
terje.kvisteroey@sensonor.no
hj@hive.no

Christian Vieider, Stanley Wissmar, Per Ericsson, Urban Halldin
Acreo AB
Christian.Vieider@acreo.se
Stanley.Wissmar@acreo.se
Per.Ericsson@acreo.se
Urban.Halldin@acreo.se

Frank Niklaus, Fredrik Forsberg, Göran Stemme
Royal Institute of Technology
frank.niklaus@ee.kth.se
fredrik.forsberg@ee.kth.se
goran.stemme@ee.kth.se

Jan-Erik Källhammer, Håkan Pettersson, Dick Eriksson
Autoliv Research
Wallentinvägen 22
SE44783 Vårgårda
Sweden
Jan-Erik.Kallhammer@autoliv.com
Hakan.Pettersson@autoliv.com
Dick.Eriksson@autoliv.com

John Franks, Jan VanNylen, Hans Vercammen
Umicore Electro-Optic Materials
John.Franks@umicore.com
Jan.VanNylen@umicore.com
Hans.Vercammen@umicore.com

Annick VanHulsel
Vito
annick.vanhulsel@vito.be

Keywords: PIMS, bolometer, MEMS, wafer scale, uncooled

New MEMS Timing References for Automotive Applications

M. Lutz, J. McDonald, P. Gupta, A. Partridge, C. Dimpel, K. Petersen,
SiTime Corporation
R. Grace, Roger Grace Associates

Abstract

Microelectromechanical Systems (MEMS) resonators have been investigated for over forty years but have never delivered the high performance and low cost required of commercial oscillators. Limiting factors have been temperature drift, long term stability, susceptibility to vibration, manufacturability, and low cost back end packaging. These requirements have now been met by SiTime's MEMS First™ wafer-level encapsulation technology combined with a system architecture using modern phase lock loop (PLL) technology combined with a high performance CMOS signal conditioning chip with integral temperature sensor to compensate for frequency offset and temperature drift. SiTime's MEMS and CMOS chips are manufactured in state of the art CMOS fabrication facilities and can be packaged with any semiconductor packaging technology such as plastic molding, flip chip, MCM and SIP. The resulting silicon MEMS oscillators outperform existing quartz solutions in various aspects such as reliability, robustness, small size, and system integration; and are suitable for most automotive and commercial applications.

1 Introduction

Electronic components in vehicles are increasing in nearly all functions including: safety systems, engine control, chassis, comfort/convenience functions, and car multi media. Nearly every electronic device needs a clock reference to operate. The typical MY 2006 vehicle, has between 15 and 50 Quartz-based time references. Crystal resonators and oscillators have been used for this application since the mid 1940's because of their high accuracy and low phase noise. MEMS researchers have been working over forty years [1-3] to develop resonators and drive circuits that can meet the stringent performance and cost specifications required to displace the incumbent quartz technology. This goal has now been achieved and MEMS oscillators which are suitable for a wide range of commercial and industrial applications are now available to address these many automotive applications. These MEMS oscillators have lower manufacturing costs compared to the quartz devices they replace because they benefit from sharing the economies of scale of front end and back end IC production. In addition the improved robustness, quality and reliability and smaller size will enable rapid adoption into the automotive market in applications such as tire pressure monitoring systems. Currently there are several applications in discussion for the initial introduction into the automotive marked such as clocks for safety critical sensors communicating by CAN and RF transceiver for tire pressure monitoring systems. First introduction is being planned for 209 after the automotive typical design-in and qualification cycle of 2 years. The first true "Single Chip Modules", such as microcontroller and communication nodes on bus systems without the need of any additional discrete components will be possible in the near future. The SiTime SiT8002 oscillator described in this paper includes a MEMS resonator die and CMOS oscillator chip in a stacked die package. The SiT8002 is the world's smallest programmable oscillator available today. The oscillator package measures 2.0 x 2.5 x 0.85 mm and resonator die 0.8 x 1.6 x 0.15 mm.

2 Manufacturing

The Quartz $3 billion industry has refined and optimized over the last 50 years its own technology to a mature level of performance, small size, and cost. On the other side the $200 billion semiconductor industry has built up one of the world largest industries supplying the demand of electronics. There are only a few functions left which can not be supplied by this industry, such as stable timing references. One of the largest barriers for commercialization of silicon MEMS oscillators and resonators has been developing a cost effective and robust packaging method. To ensure long-term stability, the MEMS resonators must be hermetically packaged in a contaminant-free environment. The cost effectiveness must be ensured by the wafer process avoiding special MEMS processes such as wafer bonding [7, 9], requiring a lot of wafer area and enabling standard semiconductor back end packaging processes such as plastic molding, flip chip, and multi chip modules.

2.1 MEMS Resonator Fabrication

The Fabrication of a MEMS First™ resonator is shown in Fig. 1. A 10-20 µm Silicon On Insulator (SOI) substrate is patterned with a deep reactive ion etch DRIE etch into a resonant structure. (2) An oxide is deposited and patterned to cover selected parts of the resonator while providing electrical contacts to drive and sense electrodes. (3) A 1.5 μm epi silicon layer is deposited and patterned with vents to access to the oxide. (4) The oxide over and under the structures to be mechanically freed is removed through the vents. (5) The resonator chamber is sealed in an epi environment with SiTime's EpiSeal process, giving an extremely clean enclosure. (6) The wafer surface is Chemical Mechanical Polish (CMP) planarized and patterned with contact isolation trenches, yielding a 10-20 μm thick mechanically durable epipoly encapsulation. (7) Insulation oxide, metal interconnections, and a scratch mask are deposited and patterned or CMOS is fabricated. The first and second silicon depositions grow polycrystalline silicon on the oxide and single crystal silicon where the oxide has been removed.

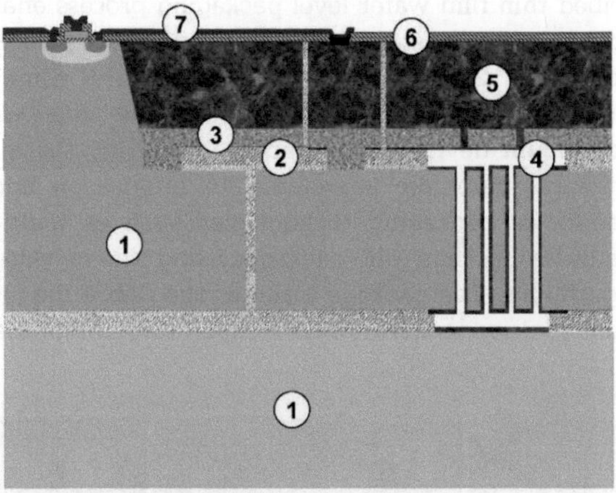

Fig. 1. Fabrication cross section.

Fig. 2. SEM of cleaved wafer before polishing of the and metal interconnect fabrication.

Fig. 2 shows a cleaved MEMS wafer after epitaxial deposition of the capping layer and before chemical mechanical polishing (CMP). The field areas are thus smooth single crystal silicon and can support fully integrated CMOS electronics. The vacuμm in the enclosed cavity has a pressure of approximately 10 mT [5, 8] and is virtually free from water and other high vapour-pressure contaminants.

2.2 Packaging Process

The MEMS die, which physically is like a standard CMOS chip, manufactured in above described thin film wafer level packaging process enables the usage of the full portfolio of well established and very robust packaging technologies developed by the semiconductor industry [6]. QFN (Quad Flat No-lead) lead frame design rules allow a pin compatible construction with the typical ceramic surface mount device (SMD) packages used by the quartz industry. Nevertheless the resonator die is suitable for integration into most of the known semiconductor packaging technologies such as multi-chip modules (MCM) and system-in-package (SIP) packages using side by side, chip stack, or flip chip configurations. The package includes the CMOS die (described later) which is needed to drive, calibrate and temperature compensate the MEMS resonator.

Fig. 3. Packaging Sequence

The packaging process as shown in Fig. 3 starts with an electroplated copper lead frame (1) on which the CMOS die (2) is attached. The MEMS die is stacked on top of the CMOS die (3). Both dies are thinned to enable packages as thin as 500 µm- 850 µm. Electrical connection is achieved by gold wire bonding (4) between the MEMS die, CMOS die and lead frame. Plastic injection molding and dicing finishes the packaging process. The chip on lead technology enables near chip scale packages which makes the SiT8002 the world smallest programmable oscillator. The smallest available footprint is 2.5 mm x 2.0 mm.

3 System and Performance

The oscillator/resonator market for timing reference is divided into 3 market segments. Low end resonators are typically used in watches, and large consumer applications such as PCs and TVs. The performance varies between 50 ppm and several hundred ppm for frequency stability. These products are relatively large and typically packaged in metal cans. The high end is represented by ultra stable and low phase noise time references such as TCXO and oven controlled oscillators used for high end wireless and communication applications. In the middle are most of the consumer electronic products such as cell phones, digital cameras, MP3 players, hard disk drives and most of the automotive applications for serial communication, microcontroller, CPUs, infotainment and navigation which all can be addressed by the MEMS based oscillator products described in this paper.

The SiTime SiT8002 is a fully programmable oscillator that may be programmed to any frequency between 1 and 125 MHz with 2 ppm resolution with a ±50 ppm tolerance stable over -40°C to +85°C. Therefore, all the common automotive frequencies are available now from a single product: for example; 8 MHz, 12 MHz, 24.576 MHz, and 66 MHz. This architecture covers all 173 industry standard frequencies provided by the quartz industry.

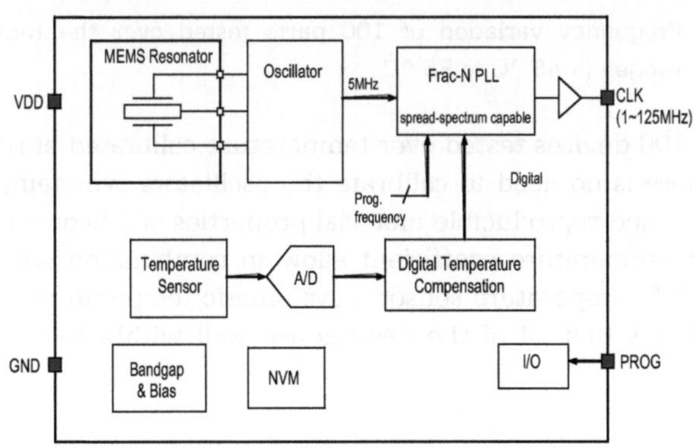

Fig. 4. System block diagram

The CMOS circuit includes a MEMS driver, temperature compensation, and a programmable frequency multiplier (Fig. 4). The power supply voltage is programmable between 1.8 V, 2.5 V and 3.3 V. This architecture can be extended for various automotive applications: The PLL can be used for wireless applications such as key-fob, tire pressure transceivers, any serial bus such as CAN, MOST and FlexRay. Future products will include spread spectrµm to reduce EMI and VCO functionality. The output drive strength is adjustable to optimize the signal for the frequency and capacitive load dependent configurations. In addition the discrete MEMS resonator is available to be integrated into customers own ASICs. The required circuit IP to drive and temperature compensate the MEMS resonator can be easily integrated onto any custom ASIC or RF-transceiver.

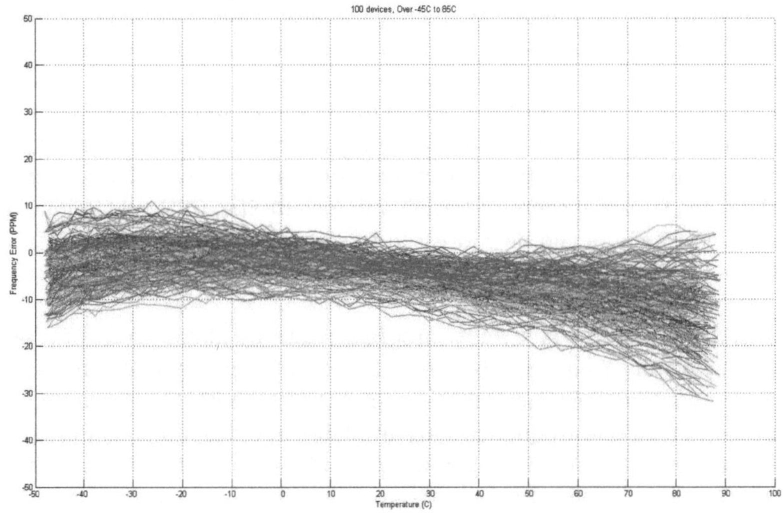

Fig. 5. Frequency variation of 100 parts tested over the temperature range of - 45 °C to 85 °C

Fig. 5 shows 100 devices tested over temperature calibrated at room temperature only. There is no need to calibrate the oscillators over temperature. The well controlled and reproducible material properties of Silicon such as Young's Modulus and temperature coefficient allow in combination with a high performance CMOS temperature sensor a systematic temperature compensation. The graph shows that all of the devices are well within the specification of ± 50 ppm.

Test Temperature	*SiT8002 typical initial frequency offset*	*Typical Quartz oscillator frequency offset*
85°C	*±20 ppm*	*±20 ppm*
100°C	*±25 ppm*	*±40 to 70 ppm*
125°C	*±35 ppm*	*±150 to 250 ppm*

Tab. 1. Comparison of frequency offset at elevated temperatures between Quartz and SiT8002 oscillators

The qualification of the extended automotive temperature range will be concluded in the near future. However, there is considerable data taken that shows excellent performance at high temperatures. For example, AT cut quartz has a very steep frequency vs. temperature curve above 85 °C, whereas silicon's frequency response remains extremely linear to temperatures as high as 400 °C. Silicon's linearity makes compensation straight forward and reliable. Tab. 1 shows initial results that MEMS First™ oscillators will outperform typical quartz based solutions.

4 Reliability and Quality

A key measure for a high performance frequency reference is drift [4]. The SiT8002 passes JEDEC standard qualification and outperforms competitive solutions. Qualification tests such as HAST (Highly accelerated stress testing: 130 °C, 85% humidity, V_{dd} +10%, 96 h), HTOL (High temperature operating life: 125 °C, V_{dd} +10%, 1000 h), HTS (high temperature storage 125 °C, unbiased, 1000 h) have been passed without failures.

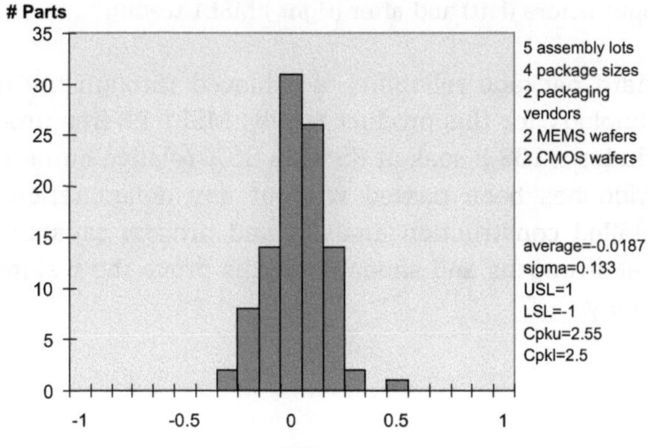

Fig. 6. Aging histogram frequency drift in ppm/year at 25 °C

Fig. 6 shows typical aging data of several lots of MEMS and CMOS wafers in different package sizes and multiple packaging vendors. Typical aging data from Quartz based oscillators are 10 times worse compared to the SiT8002 oscillators.

The MEMS First™ resonators are manufactured and sealed at 1100 °C, thus the extreme temperatures present in the automotive environment have no meaningful reliability impact on the resonator. This high-temperature manufacturing process also yields a resonator which is perfectly annealed and all mechanical stresses caused by the manufacturing are eliminated. The high temperature epitaxi sealing process removes contaminants such as water vapor and organic residues which could lead to mass-loading of the resonator and thus cause frequency drift. Known competitive packaging solutions such as low temperature wafer bonding [7, 9] or ceramic packages can not achieve this level of cleanliness which is needed for ultra small resonators where mass-loading causes significant frequency drift. Quartz and competitive MEMS solutions cannot be annealed and cleaned at such high temperatures and therefore require additional effort such as getters [9] to achieve similar results.

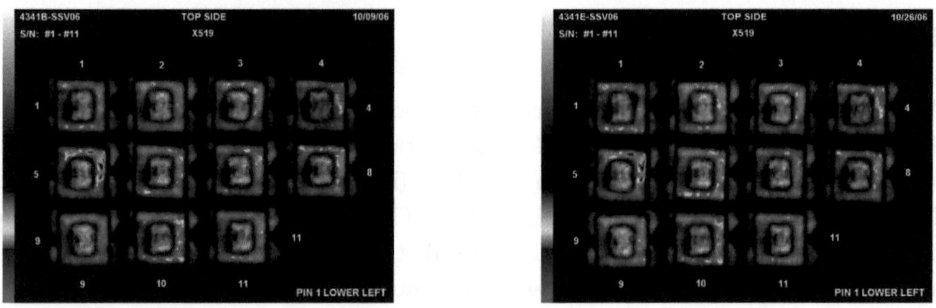

Fig. 7. Scanning acoustic microscopy image of parts in 2520 QFN packages before (left) and after (right) MSL1 testing

Industry standard package reliability is achieved through the use of mature packaging technology for this product family. MSL1 PB-free process qualification which includes a 168 h soak at 85°C in 85% relative humidity and 260°C reflow simulation has been passed without any delamination as shown in Fig. 7. The detailed construction analysis and process capabilities as shown in Fig. 8 for pull strengths and sheer strengths prove the 6 sigma design and process philosophy.

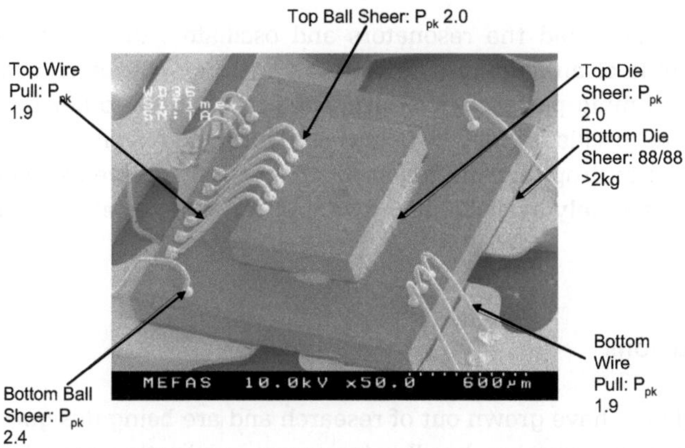

Fig. 8. Construction analysis and process capabilities

The MEMS First™ resonators also excel in their immunity to shock and vibration. SiTime tests, independent test labs, and numerous MEMS research papers [10 - 12] confirm that MEMS First™ resonators can survive shock effects that exceed 30,000 G's. No failures related to shock events or vibration at frequencies below 20 kHz have been witnessed to date. The mechanical models of the resonators show mechanical failure occurring at 250,000 G's and higher.

Test	*Description*	*SiT8002*	*Quartz oscillators from 2 different companies*
Free fall	*1m drop on steel plate 3 times same axis*	*<1 ppm*	*2-16 ppm*
Free fall	*1m drop on wood board 3 times same axis*	*<1 ppm*	*2-8 ppm*
Free fall	*1 m drop on steel plate 100 times same axis*	*<2 ppm, average variation of <0.5 ppm*	*4-30 ppm*
Mechanical shock	*1000 g all axis 3 times*	*<1 ppm*	*5-16 ppm*
Mechanical shock	*3000 g all axis 3 times*	*<1 ppm*	*5-25 ppm*
Thermal shock	*1000 cycles, - 40°C to 100°C*	*<5 ppm*	*<5 ppm*
Thermal shock MEMS resonator only	*1000 cycles, -40°C to 85°C*	*<0.2 ppm*	*na*

Tab. 2. Summary of measured temperature and mechanical shock data for MEMS First™ resonator and SiT8002

SiTime has also tested the resonators and oscillators under the shock and vibration profile of an automotive tire tread with no failures and remarkably no change in timing performance during the test. On the other hand, quartz resonators and oscillators devices show variations between ±16 ppm under the standard 1 m drop test and fail at much lower shock events. Quartz shock test data is not widely available and few application notes exist on the issue

5 Conclusions

MEMS oscillators have grown out of research and are being designed into real world applications. The key hurdles for commercialization have been solved by the technology described in this paper: frequency temperature stability, long term drift, robustness, manufacturability, cost, and integration. The basis for this achievement is the high temperature wafer level sealing method of the MEMS First™ resonator which results in a reproducible, ultra robust and stable frequency reference. Quality, reliability and performance enhancements will enable a fast penetration of MEMS First™ resonators and oscillators in the automotive and consumer market.

References

[1] C. Nathanson, R.A. Wickstrom, "A resonant-gate silicon surface transistor with high-Q bandpass properties," IEEE Applied. Physics. Letters, v.7, pp.84-86,1965.

[2] H.C. Nathanson, W.E. Newell, R.A. Wickstrom, J.R. Davis Jr., "The Resonant Gate Transistor," IEEE Trans. Electron Devices, Vol.ED-14, pp.117-133, 1967

[3] Many potential references, see for example: J. Wang, J.E. Butler, T. Feygelson, and C. T.-C. Nguyen, "1.51-GHz Polydiamond Micromechanical Disk Resonator with Impedance-Mismatched Isolating Support," Proceedings, 17th Int. IEEE Micro Electro Mechanical Systems Conf., Maastricht, pp. 641-644, 2004

[4] B. Kim, R.N. Candler, M. Hopcroft, M. Agarwal, W.T. Park, T.W. Kenny, "Frequency Stability of Wafer-Scale Encapsulated MEMS Resonators", Transducers '05, pp.1965-1968, 2005.

[5] R. N. Candler, M. Hopcroft, B. Kim, W.-T. Park, R. Melamud, M. Agarwal, G. Yama, A. Partridge, M. Lutz, T. W. Kenny, "Long-Term and Accelerated Life Testing of a Novel Single-Wafer Vacuμm Encapsulation for MEMS Resonators," In Press, Journal of Microelectromechanical Systems, 2005.

[6] A. Partridge, M. Lutz, B. Kim, M. Hopcroft, R.N. Candler, T.W. Kenny, K. Petersen, M. Esashi "MEMS Resonators: Getting the Packaging Right," SEMICON-Japan, 2006.

[7] M. Lutz, W. Golderer, J. Gerstenmeier, J. Marek, B. Maihofer, S. Mahler, H. Munzel, and U. Bischof, "A precision yaw rate sensor in silicon micromachining," International Conference on Solid State Sensors and Actuators, TRANSDUCERS '97, vol. 2, pp. 847-850 vol.2, 1997.
[8] R.N. Candler, W.T. Park, M. Hopcroft, B. Kim, and T.W. Kenny, "Hydrogen Diffusion and Pressure Control of Encapsulated MEMS Resonators," Transducers '05, pp.920-923, 2005. (Also pending in IEEE Packaging).
[9] Douglas Sparks, Sonbol Massoud-Ansari, Nader Najafi, "Long-term evaluation of hermetically glass frit sealed silicon to Pyrex wafers with feedthroughs", Journal of Micromechanics and Microengineering 2005.
[10] D. Tanner, J Walraven, Helgesen, L. Irwin, F. Brown, N. Smith, N. Masters, „MEMS reliability in shock envirnments", Reliability Physics Symposium, 2000. Proceedings. 38th Annual 2000 IEEE International.
[11] V. Srikar, S. Senturia, "The reliability of microelectromechanical systems (MEMS) in shockenvironments", Journal of Micromechanical Systems IEEE/ASME, 2002
[12] Wagner U.; Franz J.; Schweiker M.; Bernhard W.; Muller-Fiedler R.; Michel B.; Paul O.; " Mechanical Reliability of MEMS-structures under shock load", Microelectronics Reliability, Volume 41, Number 9, September 2001, pp. 1657-1662(6), Elsevier

Markus Lutz, John McDonald, Pavan Gupta, Aaron Partridge, Courtney Dimpel, Roger Grace, Kurt Petersen
SiTime Corporation
990 Almanor Ave
Sunnyvale, CA 94085
ml@sitime.com

Keywords: MEMS, resonator, oscillator, time reference, clock, timing circuit, reliability

MEMS Gyroscopes for Automotive Applications

J. Classen, J. Frey, B. Kuhlmann, P. Ernst, Robert Bosch GmbH

Abstract

This paper reviews the product lines for automotive MEMS gyroscopes for the applications ESP®, roll-over detection for airbag systems, and navigation systems by describing their silicon micromachining technology, working principle, and packaging technology. In each of these technology areas, important progress has been made in the past decade which enabled a very broad market acceptance for MEMS gyroscope sensors. Additionally, we discuss important future trends of MEMS gyroscopes that arise due to changing system requirements and due to the impact of upcoming non-automotive applications.

1 Applications of MEMS Gyroscopes in Automotive Systems

The success story of advanced MEMS gyroscopes in automotive applications started in the 1990s with the market introduction of ESP systems. The first ESP system was released in 1995 as an option for the Mercedes SL and S class. In 1997, the Mercedes-Benz A-class adopted ESP as standard equipment. This cleared the way for a broad market acceptance of this active safety system: Already in 2003, the ESP installation rate by new car registration in Germany grew to a value above 50% (Source: Bosch). In 2006 this value even reached 75%. Today, ESP is considered as one of the most powerful life saving systems in the automotive industry. This led the American National Highway Traffic safety Administration (NHTSA) to propose a new rulemaking. The proposed new Federal motor vehicle safety standard (FMVSS) No. 126 aims to require Electronic Stability Control (ESC) systems on passenger cars, multipurpose vehicles, trucks and buses. According to NHTSA studies, the installation of ESC will reduce single-vehicle crashes of passenger cars by 34% and single vehicle crashes of sport utility vehicles (SUVs) by 59%, with a much greater reduction of rollover crashes.

ESP systems rely on angular rate sensors (yaw rate sensors) for the stabilization of the car. They detect the angular rate of the vehicle around it's vertical axis and thus create the necessary data input for the stabilization system to support the driver in keeping the vehicle on track. The first industrial ESP systems worldwide came from Robert Bosch GmbH in 1995. It made use of a vibrating metal cylinder for detecting angular rate [1] (Fig. 1a). The metal cylinder was replaced in 1998 by the introduction of the first micro-machined automotive yaw rate sensor MM1 [2] (Fig. 1b). In comparison to it's predecessor, the MM1 made a dramatic progress with respect to cost and functionality. The main advantages of this sensor are it's zero point stability, accuracy and reliability. It consists of a hybrid module carrying two MEMS chips (accelerometer and gyroscope), an ASIC for the signal evaluation of both sensor elements and some discrete components. The hybrid is covered under a metal cap with a permanent magnet that is necessary for the electromagnetic stimulation of the gyroscope oscillator.

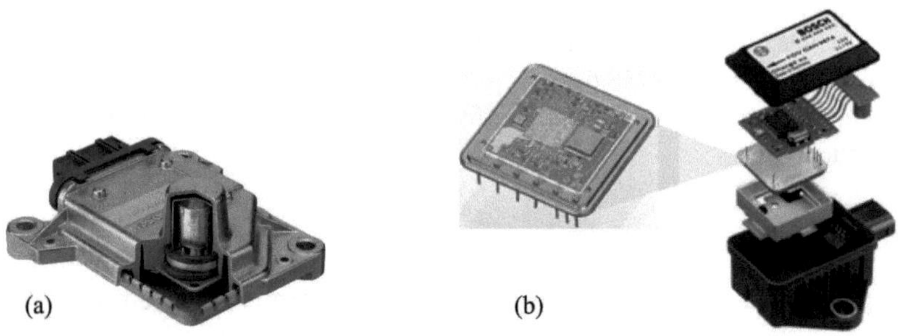

Fig. 1. (a) The first angular rate sensor of Bosch ESP® systems from 1995 (vibrating metal cylinder).

(b) Successor of (a) and first yaw rate sensor in silicon micromachining technology: DRS-MM1, introduced in the market by Bosch in 1998

In the following years, the progress is ESP systems required sensor support for additional functions. New features in advanced ESP systems include Hill Hold Control (HHC), Roll Over Mitigation (ROM), Electronic Active Steering (EAS) and the support of Four Wheel Drive (4WD), Active Suspension Control (ASC) or Steer-by-Wire (SbW). These new system functions required the development of a flexible inertial sensor cluster with a modular concept for hard- and software. This is realized with the Bosch DRS MM3 (Fig. 2) which went into series production in 2005 [3].

Fig. 2. Modular ESP sensor cluster DRS-MM3, introduced into the market in 2005

A second important field of application for advanced MEMS gyroscopes in automotive safety systems are airbag systems. In about 20% of fatal accidents, the vehicle overturns. In the case of such a rollover accident, the restraint system is required to trigger belt tensioners, side and head airbags and rollover bars in such a way as to provide the best protection possible for the vehicle occupants. The necessary sensor input for this system again comes from a gyroscope sensor that measures the angular rate of the vehicle. Other than in ESP systems, however, the sensitivity axis here is the vehicle's longitudinal axis.

Series production of a roll-over gyroscope sensor developed by Bosch [4] have already started in 2001. Since then, Bosch has released two generations of rollover gyroscopes for production [5]. Fig. 3 shows a rollover gyroscope module of the second generation. This product line is distinguished from the above mentioned ESP gyroscopes by two main features: (1) their sensitive axis is parallel to surface of the silicon chips, which allows to mount them directly on the printed circuit board of the airbag ECU, (2) they are packaged in a standard plastic full mold package, which helps to reduce their cost of production significantly.

Fig. 3. Micromachined angular rate sensor (MM2) for rollover detection systems

Gyroscope sensors as shown in (Fig. 3) are also suitable to serve in automotive navigation systems, which are the third important application of gyroscopes in automotive technology [5].

As in an ESP system, the required sensitivity axis is the vertical axis of the vehicle in order to detect the cornering of the vehicle. This sensor signal, in combination with the odometer signal, allows the navigation system to determine the position of the vehicle as a function of time exactly and thus supports the GPS navigation very effectively. It is especially important in driving situations where the GPS detector fails to provide for locating the vehicle: no or ambiguous satellite signals in parking garages, in tunnels or in the neighbourhood of tall buildings or similar (so called dead-reckoning). In general, the requirements for a gyroscope in navigations systems is very close to them of an ESP gyroscope: similar measurement range and very high offset stability.

2 Technology Progress of MEMS Gyroscopes

2.1 Silicon Micromachining Technology

The first silicon micromachined yaw rate sensor MM1 introduced in 1998 is a combination of surface and bulk micromachining. The sensor structure consists of an oscillating paddle formed out of the single crystal substrate and two accelerometers on top of it for detection of Coriolis force as described in the section below. The wafer process [2] (Fig. 4) starts on a 150 mm silicon wafer with deposition on a thick oxide layer. This oxide layer serves as an electrical isolation and sacrificial layer. Above that a doped polysilicon layer is structured providing the electrical interconnection of the MEMS structure to the outside. A second oxide layer for electrical isolation and sacrificial etching is deposited on top. The functional layer of the accelerometer is made from an epitaxial grown polysilicon layer with a thickness of 12 µm. An aluminium layer is structured on top of the functional layer giving the conductive path for driving the oscillator masses via Lorenz force, as described below and forms the electrical contacts to the external ASIC. This first process sequence is fully done within a CMOS-production line used for ASICs. In the actual micromechanics section the silicon underneath the oscillator masses is thinned down to 50 µm using KOH wet etching. The accelerometer and the oscillator masses are structured by anisotropic deep reactive ion etching, also known as the Bosch process. The MEMS structure is finally released by removing the oxide layer using HF vapour phase etching. The sensor structures are sealed on wafer level using glass frit bonding to enclose a defined atmosphere and protect the delicate device during dicing and packaging. The sensor wafer is bonded between a lower socket wafer and an upper cap wafer, as shown in Fig. 4. These two wafers are produced separately using KOH-etching. The cap wafer contains a recess above the sensor core and an opening above the external contact area to allow free movement of the sensor structures and electrical contact for measurement and packaging, respectively.

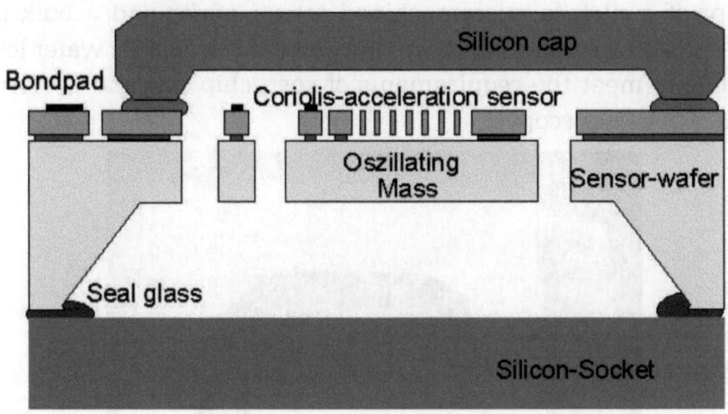

Fig. 4. Cross view of the MM1 sensor element after wafer bonding

The second type of gyroscopes [4] (Fig. 3), the MM2, uses a modified Bosch foundry process [6]. In contrast to the MM1, the element is produced purely by silicon micromachining which omits structuring of the backside of the sensor wafer. The overall microstructure for driving and sensing is made out of the same functional epipoly layer. The out of plane detection of the sensor element requires tighter control of the trench process to assure a low quadrature signal compared to the MM1. Additionally the buried polysilicon layer for electrical wiring is now used as a lower electrode to detect the capacitive change during detection. The change to surface micromachining results in a reduction of wafer cost due to a smaller number of process steps without delicate front side handling of the sensor wafer. Furthermore, the bond process could be simplified by using only one cap wafer for bonding. To achieve a sufficiently high quality factor (see section 2.2.), the design of the gyroscope requires a pressure below 5 mbar. The simplification of the wafer process and the unification with process flows used for accelerometers is a key issue to obtain smaller chip size and lower wafer costs.

The second generation of ESP gyroscopes uses a technology consisting of pure silicon surface micromachining similar to the MM2 technology. The sensor wafer is again bonded to a separate cap wafer using glass frit bonding, as seen in Fig. 5. This process generation comprises a further improvement to the accuracy of the critical MEMS processes by using high-resolution stepper lithography and tight control of the trench process reducing mechanical cross-talk. The enclosed vacuum provides a high mechanical quality factor for both drive and detection, a key to high bias stability. The improvements in wafer process resulted in a shrink of the active sensor area and provide narrow tolerance bands which are vital for large volume productions at high yield.

The concept of a surface micromachined sensor wafer and a bulk micromechanically etched cap wafer which are hermetically sealed by wafer level bonding is capable to meet the requirements of cost, chip size and performance of future generation gyroscopes.

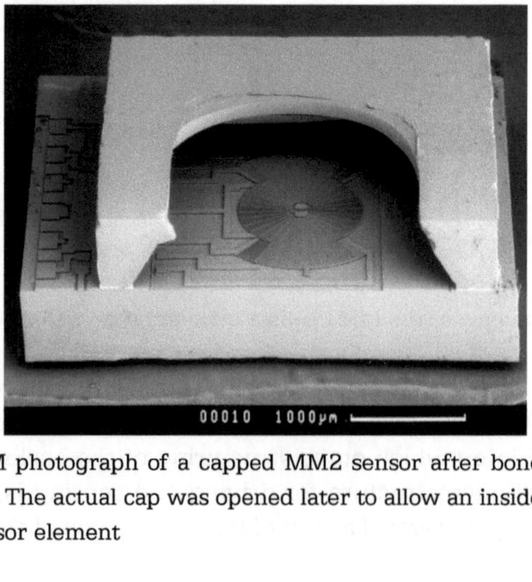

Fig. 5. SEM photograph of a capped MM2 sensor after bonding and dicing. The actual cap was opened later to allow an inside view on the sensor element

2.2 MEMS Sensor Design

All Bosch MEMS gyroscopes are Coriolis Vibratory Gyroscopes, i. e. they use the Coriolis force effect to measure the angular rate. However, the design of the different sensor elements changed considerably.

2.2.1 MM1 Design (Fig. 1b)

The principle of operation of the first generation MEMS gyroscope MM1 [2] corresponds to a tuning fork. The gyro consists of two oscillating seismic masses formed by bulk micromachining techniques and two surface micromachined capacitive accelerometers on top of the oscillating masses (Fig. 6). The measurement axis of the accelerometers is orthogonal to the direction of oscillation of the seismic masses. A rotation around the third orthogonal axis (the z-axis) imposes a Coriolis force on the accelerometers.

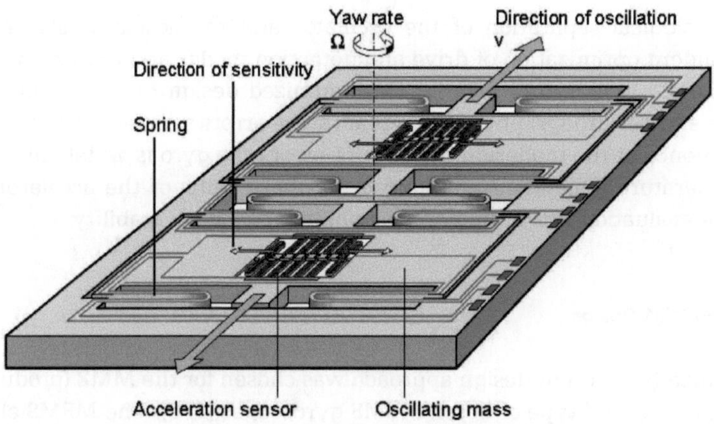

Fig. 6. Principle of operation of the MM1 sensor element

The oscillator uses an electromagnetic drive. It is mounted under a permanent magnet with a magnetic field through the chip, orthogonal to the surface, and is electromagnetically stimulated to drive amplitudes of 50 µm. The frequency of the drive mode of the original design was 2 kHz. This fairly low resonance frequency made it necessary to use an external damper attached to the housing of the device to ensure a sufficiently low level of mechanical interference (g - sensitivity) in a car even under extremely rough road conditions. To improve the g-sensitivity of the gyro the drive frequency was raised to 6 kHz by a redesign of the oscillator. By this measure the external damper became obsolete and mounting costs could be significantly reduced.

As the mass of the oscillator is fairly high (50 µm bulk silicon thickness, see above) and since there are no narrow gaps in the drive structure the Q-value of the drive mode of the 2 kHz-system is 1200 even at atmospheric pressure, high enough to stimulate the oscillator with small Lorenz forces. For the 6 kHz - system the quality factor is even higher. Hence vacuum packaging of the MEMS structures is not required.

The Coriolis force is detected with the previously mentioned surface micromachined capacitive accelerometers (Fig. 6). Comb structures which are suitable for the measurement of accelerations in the plane of the chip are employed. Closed loop position feedback is used to allow high dynamic range, large bandwidth, and good linearity of the detection circuit. The measurement of the difference between the two accelerometers filters out the linear acceleration and doubles the Coriolis signal. Synchronous demodulation using the velocity of the mechanical oscillator generates a signal which is proportional to the yaw rate.

The topological separation of the oscillator and the accelerometer enables independent optimization of drive and detection modes and reduces mechanical crosstalk (quadrature errors). The optimized design in combination with precise stepper lithography lead to quadrature errors smaller than 50°/s. As a consequence of the moderate quadrature errors the gyro is widely insensitive to temperature dependent and long term phase shifts of the accelerometers and the evaluation circuit and hence exhibits a good DC stability.

2.2.2 The MM2 Design

A completely different design approach was chosen for the MM2 (product: see Fig. 3), the second type of Bosch MEMS gyroscopes [4, 5]. The MEMS element of the MM2 is an oscillating disc with a diameter of 1600 µm. A SEM picture of the sensor element is shown in Fig. 7. The MEMS structure consists of a sensing mass, which is driven in-plane in a rotational oscillation around the z-axis at a frequency of 1.5 kHz by applying an electrostatic voltage across interdigital comb structures.

Typical amplitudes of the drive modes of MM2 are just a few degree. An angular rate Ωy around the y-axis (i.e. in the chip plane) produces a torque Mx and thus leads to an out-of-plane tilt of the oscillating disc (Fig. 8). This deflection is capacitively detected using two counter electrodes underneath the oscillating mass. As for the MM1 the capacitance changes are transformed by synchronous demodulation into an output voltage which is proportional to the angular rate. However, for the MM2 there is no position feedback, i.e., the sensor is operated in an open loop configuration which allows a simplified evaluation circuit.

Compared to the MM1 the mass of the MM2 sensor element is fairly small and the gap underneath the rotating disc is very narrow (~1.6 µm, i.e., only 0.1% of the diameter of the disk). Hence at a given pressure the quality factor of the drive mode is much smaller than for the MM1. To ensure sufficiently high quality factors of both drive and detection modes and to allow reasonably low drive voltages, a vacuum encapsulation at pressures < 5 mbar is required for the MM2 sensor element.

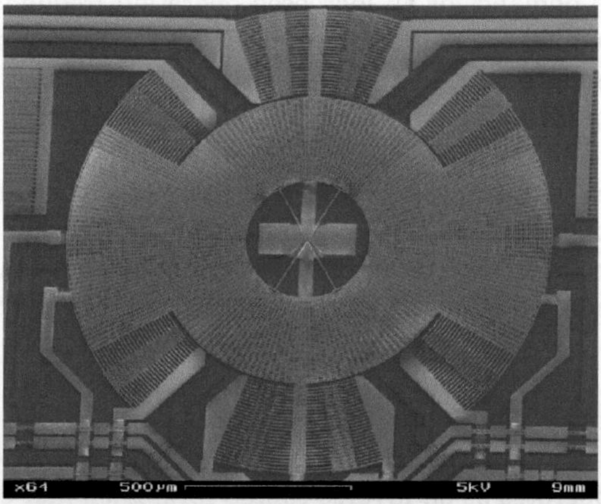

Fig. 7. SEM picture of the MM2 sensor element

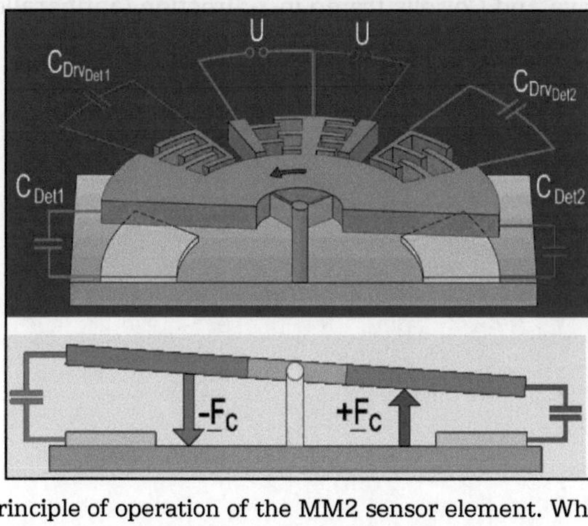

Fig. 8. Principle of operation of the MM2 sensor element. When an angular rate Ωy is applied, a torque Mx acts on the rotating disk and leads to a tilt angle. This out of plane motion is detected by electrodes underneath the movable structure.

2.2.3 Design of MM3

While surface micromachining techniques offer clear advantages in terms of size and cost, they usually implicate less signal sensitivity and therefore less accuracy and SNR. Moreover, due to the reduced size of the mechanical structures the manufacturing tolerances of surface micromachining processes tend to have an increased influence on the sensor performance. It was therefore a big challenge to develop a surface micromachined gyroscope that would provide excellent performance in terms of noise and offset stability. As the static mechanical sensitivity of Coriolis gyroscopes varies as $1/f_x$, the task was not simplified by the demand for an increased operating frequency $f_x \sim 15$ kHz. At such high frequencies sufficient suppression of mechanical interference in automotive environment can be achieved and the sensor can be readily mounted and applied without additional damping measures.

By applying new design concepts and by carefully controlling tolerances during fabrication processes of the micromechanical sensing element, those obstacles could be overcome, giving the sensor module outstanding functional performance compared to similar surface micromachined gyroscopes [4, 5, 7, 8].

The sensing element (Fig. 9, 10) is a capacitive type Coriolis vibratory gyroscope consisting of two identical spring-mass structures connected by a coupling spring. It operates as an inverse tuning fork. Each half consists of three frames (drive, Coriolis, and detection). Primary oscillation (drive) comprises movement of drive and Coriolis frame in x-direction (antiparallel mode), excited via comb drive structures.

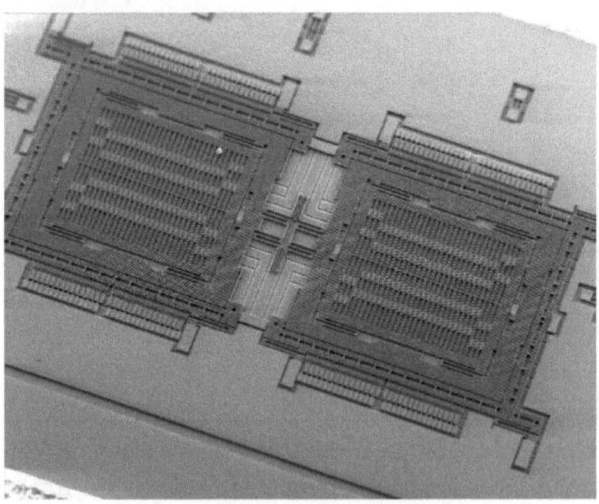

Fig. 9. SEM picture of the MM3 sensor element

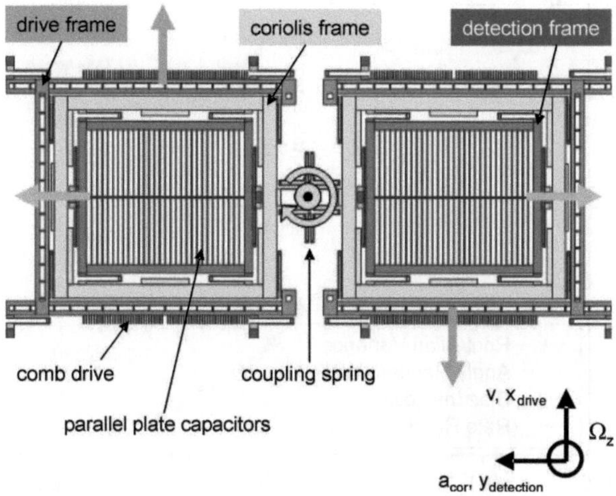

Fig. 10. Principle of operation of the MM3 sensor element

The secondary mode (detection) is excited by Coriolis forces as a result of an angular rate around the *z*-axis, it comprises oscillation of Coriolis and detection frame. Coriolis signals are detected by parallel plate capacitors within the detection frame. *U*-shaped spring design guarantees low mechanical nonlinearities. Decoupling of the detection frame from the primary oscillation reduces electrical crosstalk significantly. As mentioned in section 2.1, mechanical crosstalk is reduced by a tight control of the critical MEMS processes (e.g. lithography and trench). High quality factors for both drive and detection mode are provided by the enclosed small pressure of a few mbar.

Key component for the performance of the sensor is the electromechanical ΣΔ modulator (force rebalance control loop, including the micromechanical sensing element). By applying noise shaping design techniques the SNR of the sensor is drastically improved. The success of this sensor system approach is confirmed by experimental results [3]. The sensor module has an excellent SNR value with a typical output noise density ~0,004 $(°/s)/Hz^{1/2}$ within the pass band [0...60 Hz]. Outstanding offset stability is another key feature of the MM3 sensor module. The Root Allan Variance [9] reveals an angle random walk of typically < 14 $(°/h)/Hz^{1/2}$, and the bias instability is < 3°/h at constant temperature, see (Fig. 11). Furthermore, the rate offset deviation within the whole temperature range [-40...+120°C] is typically ±0.4°/s.

These excellent performance figures combined with high robustness and reliability as well as easy applicability of the MM3 sensor module meet the demands of present and future advanced vehicle dynamics control systems.

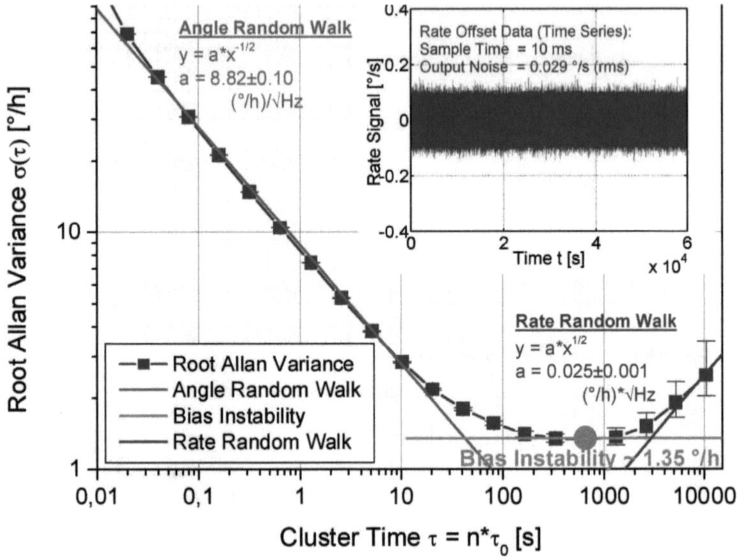

Fig. 11. Root Allan Variance of angular rate sensor SMG, and rate offset data (inset, sample time t_0 = 10 ms).

Tab. 1 provides a comprehensive overview of the generations of Bosch MEMS gyroscopes and a comparison of some key properties and parameters.

	MM1	*MM2*	*MM3*
Application	*ESP*	*ROSE, Navigation*	*ESP*
Sensitive axis	Ω_z	Ω_y	Ω_z
Package type	*Hybrid, metal housing*	*Mold*	*Premold*
Principle of operation	*Tuning fork*	*Oscillating disc*	*Inverse tuning fork*
MEMS Technology	*Combined bulk and surface micromachining*	*Surface micromachining*	*Surface micromachining*
Wafer bonding	*Glass frit (triple stack)*	*Glass frit*	*Glass frit*
Pressure	*~ 1 bar*	*< 5 mbar*	*< 5 mbar*
Drive	*Electromagnetic*	*Electrostatic (Comb Drives)*	*Electrostatic (Comb Drives)*
Detection scheme	*Closed loop*	*Open loop*	*Closed loop*
Noise [°/s/Hz$^{1/2}$] (typical)	*0.02*	*0.15*	*0.004*
Start of Production	*1998*	*2001*	*2005*

Tab. 1. Overview of design, technology and performance parameters of three types of Bosch gyroscopes

2.3 Packaging Solutions

The package of high performance MEMS gyroscopes must fulfil the following requirements:

- Unite ASIC, sensing element, and e.g. additional passive components,
- guarantee minimum misalignment of measuring element (after mounting into the ECU),
- small size,
- mounting e.g. to a PCB,
- provide electrical and mechanical connection of the IC to a calibration socket and ECU/PCB,
- suitability for labelling,
- no interference with signal to be evaluated, e.g. no mechanical stress deteriorating the signal,
- handling and environmental protection,
- excellent quality,
- high assembly yield at low cost.

These requirements are challenging to meet, which results in large development efforts. One major task is handling of mechanical stress. First of all, the detection signals are minute. Rotation of the yaw rate sensor gives rise to Coriolis acceleration which deflects the movable mass of an acceleration sensor inside the MEMS gyro. The deflection in high performance micromachined yaw rate sensor is in sub-atomic scale. Moreover, the thermal coefficient of Si (~4 ppm/K) does not match to e. g. Al_2O_3-ceramics (~7 ppm/K), copper (~18 ppm/K), mold-compound (~10 ppm/K), and LCP (~5-23 ppm/K), which results in much larger displacements and distortions as temperature changes. A quadrature signal may originate from these distortions which deteriorates the signal evaluation and may result in incorrect rate-offset values and unwanted temperature dependency. Thus even small distortions of the MEMS sensor due to thermal stress is unwanted for automotive applications like ESP and Navigation where the rate-offset at DC matters. For other applications like e.g. ROSE or consumer market, the rate value at DC is less critical. For these applications the packaging requirements are less stringent. If the thermal stress is even too large the hermetic sealing of the sensing element may crack. This is not allowed for any MEMS design.

For the first micromachined ESP yaw rate sensors (Fig. 1b), a dedicated package was developed. The micromachined sensing elements (yaw rate and acceleration) are mounted to a ceramic substrate. A ceramic substrate was chosen due to the smaller thermal mismatch, the need for electrical connection of passive components, excellent quality, and fabrication experience. The sensing elements are fixed with epoxy glue to the ceramic substrate. A soft and flexible glue is used to fix the ceramic substrate to the base plate of the metal housing. The module is assembled in a dedicated production line.

Driven by price pressure for yaw rate sensors and the high contribution of package to overall cost there is a need to reduce packaging costs without deteriorating performance. The packaging costs can be reduced using standardized and less expensive packaging technologies. This becomes possible with robustness improvements of the sensor element, signal evaluation circuit and a profound understanding of material properties. But even today stress is an issue for sensors especially where DC-offset matters. For this reason, the second generation ESP gyroscope is mounted into open cavity packages (Fig. 2) where mechanical stress is lower compared to overmolded packages. Minimizing stress in open cavity packages is done by choice of a flexible glue fixing the sensor element to the plastic housing. Overmolded packages – on the other side – are today used for e.g. ROSE sensors (Fig. 3) where rate offset at DC is less critical.

This development is ongoing. Sensor assembly is least expensive when fabrication can be done in standard production lines where high volume ICs are packaged. Therefore, the challenge of future packaging developments is to use overmolded packages (SOIC, QFN) even for high performance yaw rate sensors.

3 Outlook on Future Developments

In the past, MEMS gyroscopes have been applied predominantly in automotive applications. As described before, the technology progress in chip, design, and packaging technologies that was driven by this market has led to dramatic size and cost reductions. Both factors will allow to use MEMS gyroscopes in the future also in consumer products. Possible fields of application include portable navigation systems, image stabilization systems and gaming. These markets, however, are extremely price sensitive. As prices of devices go down, market opportunities with very high volumes appear. In many cases, handheld devices will need two- or even three-axis gyroscopes. It is expected that first products of this kind will come into the market in the near future.

Apart from these new markets, the progress in automotive systems will of course continue to create technical demands for MEMS gyroscopes. Two major requirements can clearly be seen:

- Further reduction of size of the sensor elements, the ASICs and, in consequence, the packages.
- The integration of several inertial sensor axes in one device (multi-axes gyroscopes, or combination of accelerometer and gyroscope functionalities in a single device).

Both developments are a consequence of the fact that automotive ECU's will continue to be shrunk in size, but nevertheless will integrate an increasing number of functions.

In the future, the borders between the well established vehicle domains will change. As an example, Bosch is networking the active and passive safety systems with the driver assistance systems to create the modular safety system CAPS (Combined Active and Passive Safety). The first CAPS functions, such as the predictive brake assistant and the predictive collision warning, can already be found in series-produced cars. Future CAPS functions will provide even more protection to the passenger by combining the airbag control unit with other systems and sensors in the vehicle. Advanced Rollover Sensing, for example, optimizes the protective benefits of the restraint systems when the vehicle rolls over, by including the ESP® data about the state of the vehicle in the airbag control unit's calculations. This allows the best times to trigger the belt tighteners, side and head airbags and the roll bar to be determined even earlier and more reliably. Moreover, data from navigation systems will be included in future safety concepts. Information about the car's current position and the layout of the roads can, for instance, be used to warn the driver in advance of danger spots, e.g. dangerous crossings, tight curves or accident blackspots.

As a consequence of this fusion on system level, the inertial sensors of the according systems will have to merge too: Since the data from vehicle dynamics sensors such as gyroscopes and accelerometers will be used by numerous systems, it is obviously a good idea to integrate them in a multi-axes inertial measurement unit. The development of such a device, however, will require much more effort of the kind described in the sections above: Progress in MEMS technology, along with advanced design and packaging will continue to be the success factor for further reduction in cost and increases in customer benefit.

References

[1] "Yaw Rate Sensor for Vehicle Dynamics Control Systems", A. Reppich, R. Willig, SAE Technical Paper 950537 (1995).

[2] "A Precision Yaw Rate Sensor in Silicon Micromachining ", M. Lutz, W. Golderer, J. Gerstenmeier, J. Marek, B. Maihöfer, S. Mahler, H. Münzel, U. Bischof, in Proceedings of Transducers '97, Chicago, IL, June 1997, p. 847-850.

[3] "New Surface Micromachined Angular Sensor for Vehicle Stabilizing Systems in Automotive Applications", U.-M. Gomez, B. Kuhlmann, J. Classen, W. Bauer, C. Lang, M. Veith, E. Esch, J. Frey, F. Grabmaier, K. Offterdinger, T. Raab, H.-J. Willig, R. Neul, Proceedings of Transducers'05, Korea, Seoul, June 2005, pp 184-187.

[4] "A Low Cost Angular Rate Sensor in Si-Surface Micromachining Technology for Automotive Application", A. Thomae, R.Schellin, M. Lang, W. Bauer, J. Mohaupt, G. Bischopink, L. Tanten, H. Baumann, H. Emmerich, S. Pinter, K. Funk, G. Lorenz, R. Neul, J. Marek, , SAE 1999, Detroit, Michigan, USA, March 1999, Technical Paper 1999-01-0931.

[5] "Bosch Angular Rate Sensors – Advanced Sensor Technology for Innovative Applications", M. Keim, M. Lang, B. Breitmaier, M. Grossmann, S. Zunft, J. Classen, M. Koc, M. Veith, G. Wucher, M. Offenberg, E. Steiger, T. Lich, G. Noetzel, in Proceedings of COMS 2003 (Commercialization of Microsystems), Sept. 2003, Amsterdam (NL).

[6] M. Offenberg, F. Lärmer, B. Elsner, H. Münzel, W. Rietmüller, Novel Process for a Monolithic Integrated Accelerometer, Transducers '95, Stockholm, June 1995, pp. 589-592.

[7] W. A. Clark, R.T. Howe, R. Horowitz, in 7th Solid-State Sensor and Actuator Workshop, Hilton Head Island, S.C., June 1996, p. 299-302.

[8] W. Geiger, J. Merz, T. Fischer, B. Folkmer, H. Sandmaier, W. Lang, in Proceedings of Transducers '99, Sendai, Japan, June 7-10, 1999, p. 1578-1581.

[9] J. A. Geen, S. J. Sherman, J. F. Chang, S. R. Lewis, IEEE Journal of Solid-State Circuits, Vol. 37, No. 12, 2002, p. 1860-1866.

J. Classen, J. Frey, B. Kuhlmann, P. Ernst
Robert Bosch GmbH
Tuebinger Strasse 123
72762 Reutlingen
Germany
peter.ernst2@de.bosch.com
ralf.nossek@de.bosch.com

Keywords: MEMS, automotive sensors, gyroscopes, Bosch, surface micromachining, packaging, ESP, rollover sensing, navigation systems, active and passive safety systems, bulk micromachining

Silicon Technology enabling Cost effective HAR Structures

M. Tilli, Okmetic Oyj

Abstract

Rapidly developing silicon-based sensor technology also necessitates new material solutions in the field of silicon wafers. Combined with the constantly accelerating product cycles, this means that new sensors need to be launched cost-effectively and quickly. Collaboration between the material supplier and the customer must be seamless and continuous in order to be able to avoid unnecessary business risks. Traditional sensor technologies are now joined by high aspect ratio (HAR) structures that enable the products' cost-effectiveness to be improved even further, as the method makes allowances for small sensors and large wafer sizes, and is mostly compatible with the technology used in the manufacture of semiconductors. HAR technology also enables full advantage to be taken of the possibilities presented by SOI: in the manufacture of SOI wafers, it is possible to introduce tailored buried structures in the wafer. This opens up new vistas in terms of manufacturing high-performance sensors that integrate multiple functions. This presentation focuses on SOI technology that enables HAR structures with DRIE processing.

1 Silicon as a Raw Material in Sensor Manufacture

Silicon will maintain its position as the primary raw material in sensor manufacture, thanks to its long-lasting mechanical and chemical stability, its predictable elastic behaviour and the possibility of making use of its anisotropic properties. Moreover, silicon's long history in the manufacture of semiconductors brings synergy to sensor manufacture. New sensor manufacturing techniques, DRIE and vapor phase etching of the sacrificial layers present opportunities for reducing the sensors' size even further without compromising performance and reliability. The industrial sector's desire to integrate as many functions as possible into a small space favors new silicon material solutions. The new kinds of silicon wafers can, for example, be used to make inertia sensors with several axes of sensitivity (3D acceleration, 3D gyro), and on the same piece of silicon can be integrated an pressure sensor, for example, in addition to the inertial sensor. The silicon wafer can be used as a platform for a 3D magnetic sensor and, if necessary, the sensor and electronics can be integrated on the same piece. No other material offers the same level of freedom.

The automotive industry's aspiration to add more and more functions into cars in a cost-effective manner by building sensor clusters is now possible thanks to optimization based on new silicon material solutions. Using means learned from semiconductor technology it is possible to make small sensor modules in which a semiconductor piece containing electronics and the sensors themselves are stacked onto each other and thinned. While the new technology evolves, new business models are springing up within the industrial sector, as the lines between the responsibilities of the material supplier, the sensor manufacturer and the system integrator become more and more blurred. Just as the sensor manufacturer and the system integrator have to work closely together, so too the new manufacturing techniques increasingly call for cooperation and customizing the material to fit the sensor manufacturer's process. In some cases, customization can even be sensor-specific. Certain structures that are critical to the sensor have to be constructed as early on as in the manufacture of the raw material, the silicon wafer, which means taking into account the special requirements that the sensor sets in terms of the material.

2 HAR Structures in Sensors

One of the biggest challenges in sensor technology at the moment relates to achieving cost-effectiveness and being able to integrate several functions on a single piece of silicon without sacrificing reliability and performance. Once the sensors' production volumes reach larger scales, cost-effectiveness can be improved in many ways. Reducing the size of the sensor has a direct correlation to a lowering of unit costs. Increasing the wafer size increases the number of sensors that can be built on the wafer, while the cost of processing the wafer stays essentially the same as with smaller wafers provided that the production line's capacity is used to a satisfactory degree. The performance and reliability of sensors made using traditional wet etching techniques may be good but reducing the sensor's size is hardly straightforward. In order to be cost-effective, wet etching often requires the use of extremely thin wafers, which makes increasing the wafer size problematic.

The limitations of traditional sensor manufacturing technologies can be overcome by using high aspect ratio (HAR) structures, and HAR structures currently represent the best way to reduce the die size. Unlike in anisotropic wet etching, the entire surface area of the piece of silicon can be made use of, as the structures are separated by narrow RIE-etched trenches. If an SOI wafer is used as the raw material, the thickness of the structures can be controlled very accurately. Moreover, the buried oxide in SOI wafers can be etched away, if necessary, which also makes the structure mobile when detached from the platform. Fig. 1 shows a typical detail in a HAR structure. The technology enables not only the cost-effective manufacture of large volumes of sensors but also the development of even more diverse sensor structures. The wafer size can be increased almost unlimitedly and, with certain restrictions, technology based on HAR structures is mostly compatible with the CMOS line and only requires a few special processing devices – as a result, economies of scale can be achieved by using the line for both semiconductor and sensor wafers, if necessary.

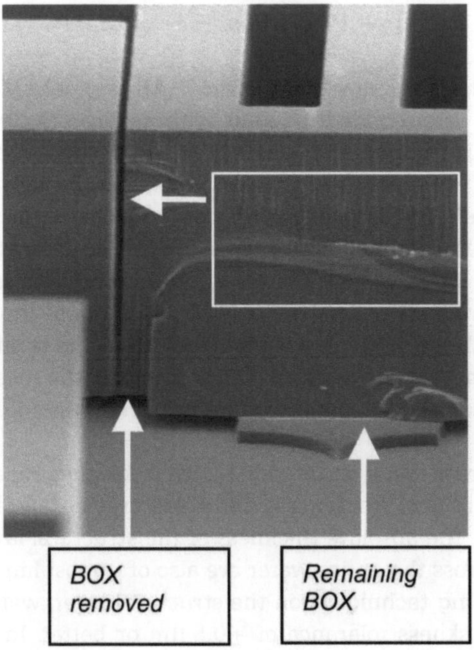

Fig. 1. Detailed view of a HAR structure once it has been etched into an SOI wafer. [1]

The technology can be used, for example, for making 3D accelerometers. Moreover, the technique allows both the IC and the sensor to be integrated on the same piece of silicon. Kiihamäki et al [2] have described a method where at first a sensor is pre-made using an SOI wafer, the structure is then sealed and planarized, and a CMOS is finally produced. The sensor structure can then be given the finishing touches at low temperature after protecting the CMOS part. It has been demonstrated that this method can be used for making integrated pressure sensors, for example, and other kinds of structures as well.

This technique can also be used to create anti-stick structures for the moving parts of the sensor, in which case a narrow air gap can be used underneath the membranes, for example.

In addition, HAR structures can be used to produce extremely thin sensors which would not be possible with wet etching techniques. Once the wafer that contains the sensor element is sealed using another silicon wafer, the structure can be easily thinned using rubbing techniques borrowed from semiconductor technology.

3 SOI Wafers and Their Properties

SOI wafers are the ideal raw material for HAR technology. In an SOI wafer, two wafers have been bonded together with a silicon oxide film in between. The other wafer is thinned mechanically to the desired thickness. The thickness of the bulk wafer and the thinned wafer can be tailored on a case-by-case basis relatively freely with the typical thickness of the bulk wafer being 400-500 μm and the thinned active layer in HAR structures being 5-100 μm; the buried oxide (BOX) is typically 0.5-2 μm. If the wafer is to be used on a semiconductor production line, the overall thickness of the wafer should be chosen in line with the SEMI standards. In principle, the technique can be used to make all commercial wafer sizes, but in practice, the most common size is 150 mm while the use of 200 mm wafers is just beginning.

Since the structure is made on the thin active layer, the properties of this layer are vital. As the critical elements in the structures are usually flexible films, springs or beams, the absolute thickness of the structural layer and the thickness tolerance across the entire wafer are also of utmost importance. By using mechanical thinning techniques on the structural layer, wafer manufacturers can achieve a thickness tolerance of ± 0.5 μm or better. In other words, the measurements taken from two different wafers and across an individual wafer are within these values. It is also important to remember that the accuracy is basically independent of the layer thickness; in other words, extremely good relative accuracy can be achieved with thicker layers as well. Fig. 2 shows the distribution of thickness measurements taken from a large batch of wafers. The film thickness of each wafer has been measured from nine different points.

The active layer can be made with extreme perfection, and the junction is practically free from voids. A good quality junction is also a requirement when the oxide under the structures is to be subject to either wet etching or vapor phase etching. If the strength of the junction is insufficient, the junction will etch faster than the thermal oxide interface, compromising the controllability of the etching process.

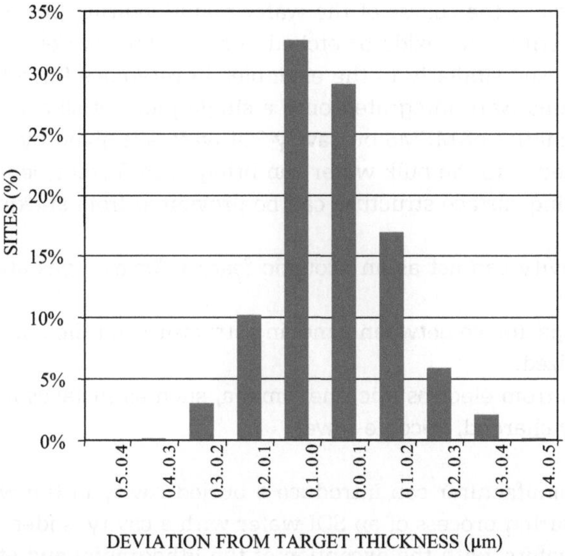

Fig. 2. Distribution of thickness measurements of the SOI wafers active layer.

SOI wafers present sensor engineers with a notable benefit, since they enable wafer manufacture where the bulk wafer and the active layer can have different crystal orientations. The bulk wafer can be (100) oriented, for example, and the active layer (110) or (111) oriented. This makes it possible to take advantage of silicon's anisotropic elasticity or piezoelectric properties. Similarly, it is possible to make SOI wafers with two thin active layers on top of each other, for example, with an oxide in between.

The resistivity values and types of the bulk wafer and the active layer can differ from each other. Resistivity can be extremely low or extremely high – from a few milli Ω/cm to over a thousand Ω/cm. High resistivity is needed, for example, for minimizing electrical losses in RF applications; good conductivity is useful for preventing structures from becoming electrically charged or when the wafer is used for transmitting a signal. It is, however, important to remember that the thickness of the active layer is measured meticulously by means of infrared interferometers, and if the resistivity is too low and the film thick, it is impossible to take a measurement and a less accurate capacitive measurement technique must be employed.

Although the efficient use of SOI wafers makes many different structures possible, in some situations further processed SOI wafers provide an interesting alternative. The wafer manufacturer can introduce a cavity buried underneath the active layer in the course of the wafer manufacturing process. The cavity can be a pattern in the oxide or etched onto the bulk wafer. A cavity in the oxide can be used similarly to the examples in reference [2], where semiconductor functions were integrated onto a single piece of silicon. A good overview of possibilities of MEMS on cavity-SOI wafers is given by Luoto et al [3]. A cavity etched onto the bulk wafer can bring even further benefits:

- A moving surface structure can be prevented from sticking to the bulk wafer.
- The cavity can act as an acoustic feature (or as a gas attenuation feature).
- The capacitance between a moving structure and the bulk wafer can be minimized.
- Effects from electrostatic phenomena, such as surfaces becoming electrically charged, become fewer.

The wafer manufacturer can introduce a buried cavity in the wafer, because the manufacturing process of an SOI wafer with a cavity is identical to that of normal SOI wafers, with the exception of the lithography and etching stages. This is why economies of scale and the resulting cost-effectiveness are attainable. The size and shape of the cavity depends on the thickness of the active layer, which means that the possibilities and limitations of SOI wafer manufacture must be taken into account when designing the sensor structure.

The same that has been stated above with regard to the possibility of making SOI wafers so that the bulk wafer and active layer have different crystal orientations also applies to pre-processed SOI wafers.

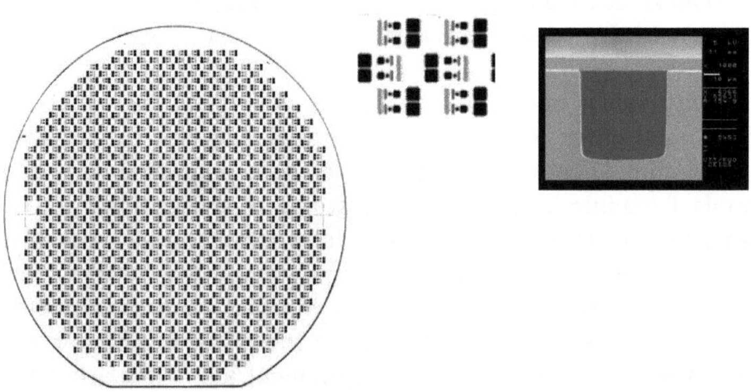

Fig. 3. SAM microscopic image of cavities made in an SOI wafer and a cross-sectional view of the cavity

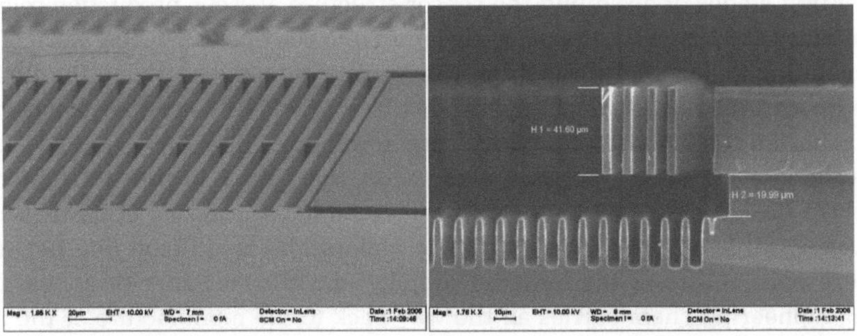

Fig. 4. Detailed view of a RF delay line made using the cavity structure [1]

4 New Approach

When manufacturing sensors and especially when using modern, highly advanced and customized wafers, seamless and close cooperation between the wafer material supplier and the sensor manufacturer is crucial.

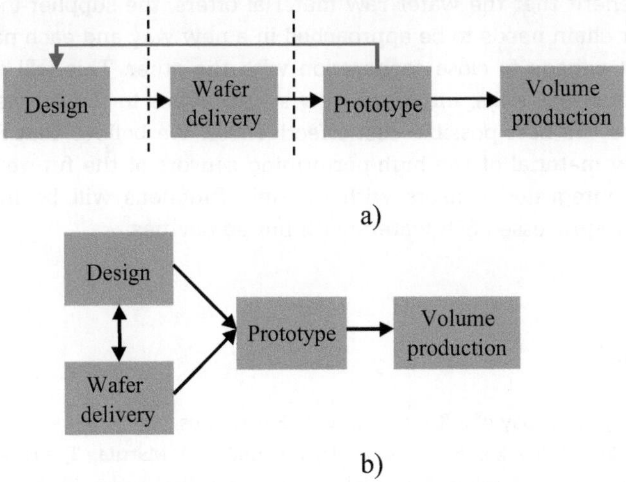

Fig. 5. The new approach within the industrial sector: the line between the supplier and the customer becomes blurred. a) Traditional way of working - information barriers between design, wafer manufacturing and sensor manufacturing, b) shortened time-to-market by removing the information barriers by early close collaboration.

When the material supplier works closely with the customer starting from the very first stages of designing the sensor structure, various production-related feasibility aspects can be taken sufficiently into consideration, therefore often eliminating delays that result from the need to amend designs. In the future, the growth in new kinds of applications in the field of consumer electronics will shorten the lifespan of products, and the shorter timescales that this kind of collaboration makes possible along with the ability to prevent material-derived design risks will be a significant advantage in terms of costs. When it comes to using pre-processed silicon wafers, the traditional line between the responsibilities of the material supplier and the customer disappears, as the engineering chain carries all the way back to the manufacture of the raw material.

5 Conclusion

Above, we have discussed the possibilities that silicon wafer development opens up in terms of sensor manufacture. New, more advanced silicon wafer materials make it possible to develop sensors that feature even better performance and cost-effectiveness as well as to integrate an even higher number of functions onto a single piece of silicon. In order to be able to realize every possible benefit that the wafer raw material offers, the supplier-to-customer-to-end-user chain needs to be approached in a new way and each party in the chain must engage in close cooperation with the other. This will allow us to minimize business risks, introduce new sensor types to the markets quickly and achieve the best possible cost-effectiveness. We believe that SOI wafers are the raw material of the high-performing sensors of the future, and especially that integrated sensors with multiple functions will be increasingly based on pre-processed SOI wafers with buried cavities.

References

[1] Photo by courstesy of J Kiihamäki, VTT Electronics

[2] Kiihamäki, J; Pekko, P; Kattelus, H; Sillanpää, T; Mattila, T, Proceedings 5th Symposium on Design Test, Integration and Packaging of MEMS/MOEMS, Mandeliou - La Napoule, FR, 5-7 May 2003,. Institute of Electrical and Electronics Engineers. France (2003), 229 - 233

[3] Luoto, H; Henttinen, K; Suni, T; Dekker, J, Mäkinen, J and Torkkeli, A, Solid State Electron (2007), in printing

M. Tilli
Okmetic Oyj
Piitie 2, 01510 Vantaa
Finland
Markku.Tilli@Okmetic.com

High-End Inclinometers - Evolution of Digital Platform towards Performance, Safety and Sensor Fusion

T. Vilenius, VTI Technologies Oy

Abstract

Bulk micromachining from VTI Technologies Oy is well known for excellent performance especially for low-g measurements. The latest development of VTI 3D - MEMS has combined higher aspect ratios reached with deep reactive ion etching with those bulk structures. This has allowed shrinking the die size of sensing elements in combination with even better performance than earlier generation of accelerometers and inclinometers. Parallel to this dedicated integrated circuits have been developed in a way that better and more precise signal conditioning and highly advanced safety features were possible in the same die. This has provided an excellent tool box for failure detection that enables latest automotive application to come along with a minimum amount of physical sensors; so called sensor fusion.

Accurate inclination measurement opens various possibilities for improved or new functionality and applications in a vehicle from Hill Start Aid through headlamp levelling to improved navigation. Improved performance of sensors facilitates usage of higher range sensors also as a high performance inclinometer, enabling sensor fusion, adding new applications without increased system cost.

This article describes the roll - out of VTIs new digital inertial platform in order to fulfil the various requirements in automotive environment with a minimum of sensing hardware by giving examples from existing functions in combination with new application design-in. Methods and environmental aspects will be explained to display that a clever shrink in MEMS comes along with higher measurement performance, more sensing options, higher reliability in terms of PPM figuress as well as system availability and self - detecting capabilities inside even most sophisticated automotive systems.

1 Family Concept

1.1 1-2-3 Axes in the same Housing

The VTI digital inertial platform covers *X*-, *Y*- and *Z*- axis and all combinations of the three in same package so that depending on system requirement the best suiting sensor is "just dropped in" to the PCB.

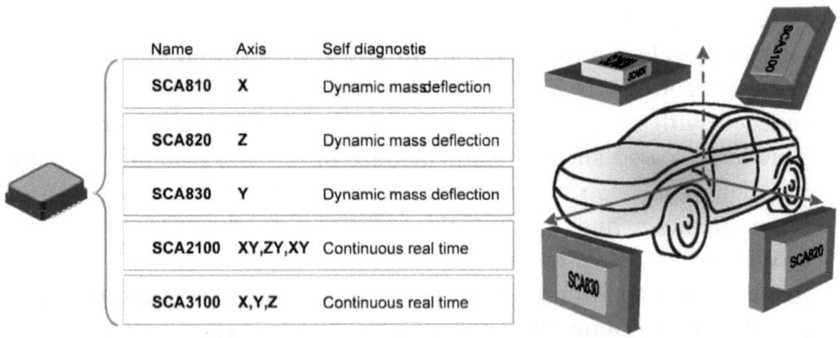

Fig. 1. Combination of sensing directions within the same housing

A mixing of existing sensor elements and ASICs with next generations is even possible to generate an easy transfer from generation to generation. This flexibility in introduction makes the change management much more reliable for the specific application. Within the same housing with uniform layout and pin - out, a variety of products will be available so that the application-tailored performance, number of axes as well as safety level can be adapted to the requirements.

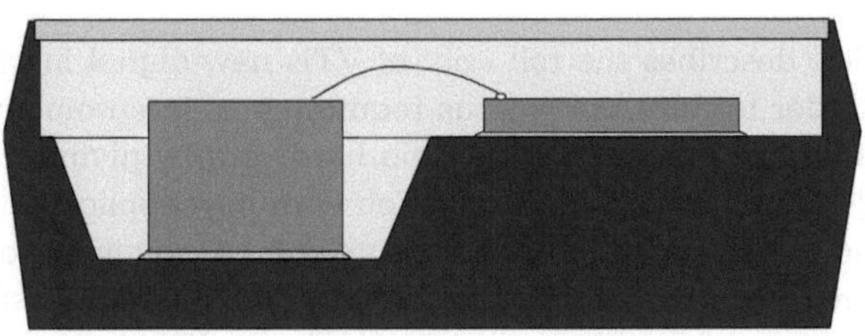

Fig. 2. Basic structure of sensing element and ASIC in premolded housing

In automotive applications temperature requirements of -40°C up to 150°C lead to extraordinary requirements in MEMS packaging. Best results were achieved with premolded packages where the CTE mismatch between housing plastic and silicon MEMS is buffered best by silicone gel and adhesives. This leads to uniform behaviour over temperature as well as negligible hysteresis effects. The disadvantage of higher costs of this more sophisticated packaging method can be compensated by stable high yield for the packaging process as well as better performance products.

The chosen Dual Flat Lead (DFL) concept is based on a lead frame premolded package. The sensing element as well as the ASIC are picked and placed by die bonder and positioned on silicone adhesive. The open cavity is filled with silicone gel and protected with a stainless steel lid, which is used for laser marking identification with product name as well as unique serial number for traceability. Both are in clear text as well as 2D - matrix code for automatic reading. The product is produced on a high performance, automated line with integrated test area.

The pins of the housing are chosen differently from SCA610 family housing before, where gull wings were considered to be required to meet the thermal shock reliability requirements. The housing for the next generation was produced as prototype in two versions – one with gull wings and one with flat leads underneath the housing. Thermal shock comparison tests with traditional tin-lead solder as well as lead free solder have shown that the results with the flat leads were fully acceptable for the harsh automotive reliability requirements. Compared to the standard QFN housing the pins realized here are having excess leads of 0.3 mm (Fig. 3). The advantage is that the surface of the pins was increased and even more important that a clear meniscus is building up during soldering process that is improving the reliability of the soldering. In addition to that the solder joint and meniscus can be inspected automatically by visual inspection.

Fig. 3. New DFL component housing (7,6 x 8,6 x 3,3 mm^3) and soldering meniscus due to 0,3 mm excess leads

The chosen housing concept is therefore dedicated to reliability requirements of harsh automotive environment. Due to the premolded housing the performance of the sensor is the same before and after soldering process. This is important to avoid any over temperature tests in application at end of line. The concept has proven to be extraordinary reliable. Components have passed 3000 temperature shocks in 2nd level reliability testing, as well as mechanical bending of the PCB for 200 000 cycles, well beyond the industry standard life time requirements.

1.2 3D - MEMS Sensing Element Technology

The single axis sensors are further developments of the existing single axis sensors, shrinking of the die size have been enabled by new process technologies without sacrificing of the active area and thus reducing the active signal. The technology is well proven and has been taken into use in the analog product during the life cycle facelifts. The high reliability is inherited from earlier generations of MEMS-elements while improving in many details.

The multi-axis element takes the MEMS sensors to completely new level. New process technologies, combined with use of SOI (Silicon On Isolator) wafers allows not only higher aspect ratios, but also facilitates usage of high precision torsional springs.

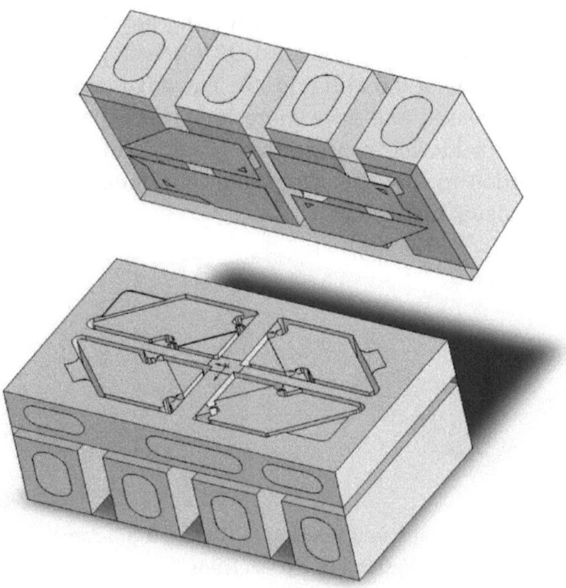

Fig. 4. 3-axes sensing element interior

The VTI 3D - MEMS structures guarantee that in - range failures and sticking are impossible due to the fact that the bulky mass structure etched into single crystal silicon doesn't have an in-range position where electrostatic adhesion might be more powerful than low acceleration forces as present in low- g applications. Single crystal silicon does not show any creeping or other deterioration over lifetime. For applications which have to work fast after power-up VTI can guarantee that there isn't any start - up drift. This is quite important for failure recognition algorithms as well as for low-power applications were the power is turned on only short time to save energy.

1.3 Signal Conditioning

The sensing element with its symmetrical dual capacitor structure for best common mode suppression is read out by advanced sigma delta conversion in an analogue interface block. This gives several degrees of freedom related to the resolution that can be varied between 8 and 16 bits depending on application demand. The low resolution comes with fast update frequency for higher cut-off frequency applications, whereas the high resolution is dedicated for slow applications like inclination measurements. A fast SPI interface with up to 8 MHz clock frequency and strong driving capabilities is realized to transfer the information to the application controller. The SPI interface is realized in a way that it has the same format for all types of sensors – single, dual and three axes – to achieve the platform character with full interchangeability of motion products within the application.

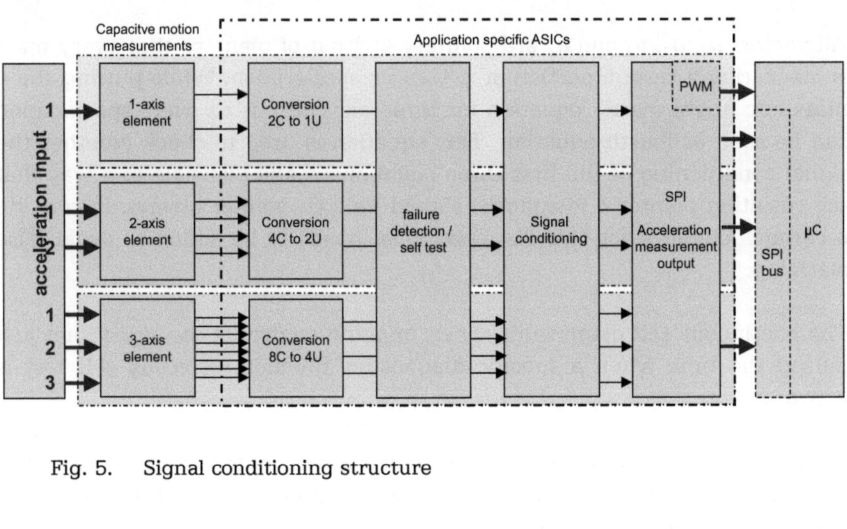

Fig. 5. Signal conditioning structure

1.4 Safety Features

A completely new level of self diagnostics was reached through the availability of redundant mass in the multi-axis sensing element.

$$STC(a_1, a_2, a_3, a_4) = (a_1 + a_3) - (a_2 + a_4) \equiv 0$$

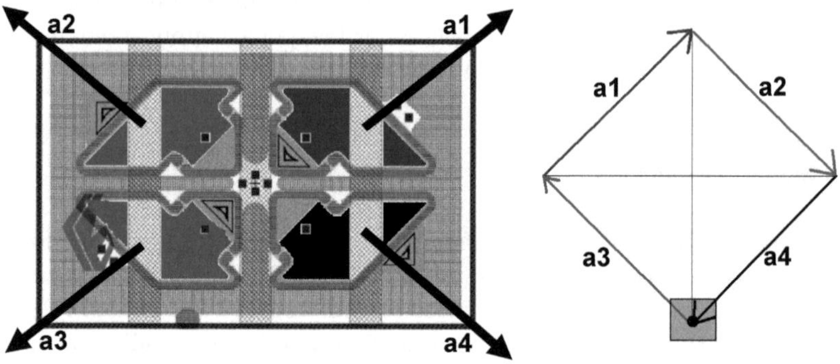

Fig. 6. Diagnosis 0-pointer of acceleration vectors as safety feature

$$\begin{pmatrix} X \\ Y \\ Z \\ d \end{pmatrix} = \begin{pmatrix} -g_{11} & g_{12} & g_{13} & -g_{14} \\ g_{21} & g_{22} & g_{23} & g_{24} \\ g_{31} & g_{32} & -g_{33} & -g_{34} \\ g_{41} & -g_{42} & g_{43} & -g_{44} \end{pmatrix} * \begin{pmatrix} a_1 \\ a_2 \\ a_3 \\ a_4 \end{pmatrix} \tag{1}$$

All vectors a_1, a_2, a_3 and a_4 are pointing 45° out of plane so that every mass is measuring a certain portion of x, y and z acceleration. While putting the 4 mass into above matrix equation for three axes output an over - specification can be seen as fourth equation. This equation is used to check whether the signal conditioning of the first three equations is still plausible in a way that the resulting pointer d (diagnosis) should be a 0 - pointer always. In real life a certain tolerance for this diagnosis level needs to be added to avoid false alarming.

The continuous self diagnostics allows ongoing control of the signal, not just during the time when a specific diagnostics function, typically self test is activated.

In the single axis sensor the proven self diagnostics method is used – the electrostatic forced self test. In this test voltage is applied to one of the capacitors deflecting the proof mass to one side and the output to rail. The self test was modified in a way that the mass can be deflected to both directions in SCA8X0 series. A control was implemented to release the mass automatically after reaching a pre-defined threshold. After successful self test the result can be read out via SPI. If the self - test fails, the sensor is recognizing it by logic so that a signal not valid signal status is set.

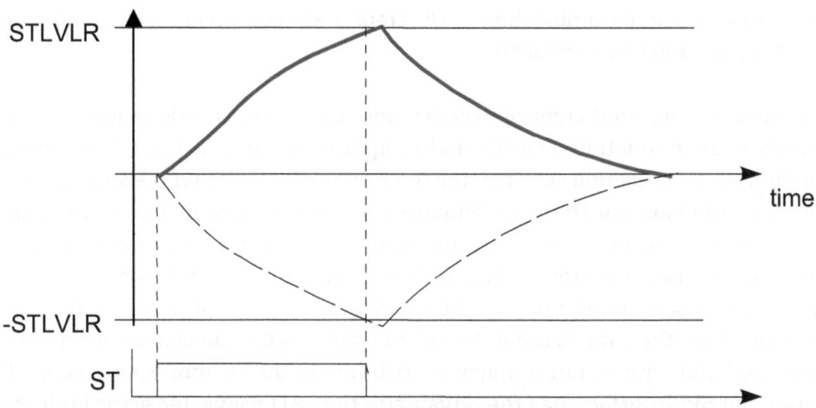

Fig. 7. Self test sensing element signal of single axis SCA800 family

The ASIC diagnosis is designed in a way that malfunctions of the sensing element as well as for the interconnections are detected. In combination with a straightforward sensor structure this leads to negligible field failure rates as well as to extremely low FIT Figures for system safety calculations.

2 Application Examples

The established applications like ESC, electrical suspension control and 4WD ABS benefit from the improved performance and failure detection. New performance level facilitates new applications like hill start assistance, electronic parking brake, satellite navigation support in multilevel roads or poor satellite receiving conditions, roll stability control, roll-over-detection, sophisticated anti-theft alarm systems with inclination and even with the detection of human being inside the car.

The multi-axis element design facilitates different g-ranges in different measurement directions. This allows integration of the sensor to various systems in the vehicle where measurement data from different axis can be used for different applications from safety critical ESC and roll over to various comfort applications like hill start assistance.

The resolution can be changed according application needs and true sensor fusion can be reached by using the same sensor as an extremely high resolution inclinometer and a 2 g accelerometer, possibly for two different applications.

2.1 Sensor Fusion: Combination of Different g-Ranges while maintaining high Measurement Performance

The change from total error of 100 mg into offset error levels required by hill start assistance function over life and temperature range is not free of charge. The tighter specification sets stricter requirements for the whole supply chain. The most obvious are the requirements for the sensor stability over lifetime, sensor manufacturer's test capability and PCB assembly accuracy. While 1° assembly error constitutes already a 17 mg error the calibration of the system offset after assembly into vehicle becomes a necessity, not just a system optimization. Also the total error in the signal chain after the sensor output is no more negligible input, but a major contributor to the system accuracy. In this matter the SPI interface has the advantage that AD converter error budget as well as ratiometric failures can be neglected on application side.

After launching all sensors in the standard configuration VTI is now heading towards combination of different g-ranges within just one multi-axes sensor. This is done by different scaling of multi-mass system so that the basic principle as shown in Fig. 6 is remaining the same. A most typical combination is a three axes sensor SCA3100 with 2g and 5g in combination as well as high offset stability. With this combination ESC, rollover, 4x4 ABS and EPB function will get their input from just one sensor with enhanced failure detection. If required a second single axis sensor can be driven on the same SPI for redundancy purpose for active steering application.

2.2 Vibration Measurement

A car is a very vibrant environment. While for standard ESC requirements a system with typically 100 mg offset stability can focus on measurement of the dynamic behaviour of the vehicle, a system for inclination measurement of e.g. 2° tilting angle accuracy – equal to 35 mg offset failure has a significant input starting from erratic signals from the entertainment system, driver and passengers of the vehicle, which might be much larger than the side wind force that should be counterbalanced by the enhanced vehicle stabilization systems. Integration of the measurement functions to existing modules around the vehicle, including ECUs located underhood mean also more mechanical noise in the measurement signal. In ideal case only the significant signal bandwidth is measured and further processed, reducing the room for misinterpretation of signals. The 3D - MEMS by VTI enables damping of the over frequency signals and thus making system activation decision easier by filtering out the entertainment and human originated signals not only through signal processing but by true mechanical filtering. Thus it can be said that in automotive environment less is truly more.

In terms of vibration measurement measuring of small signals has never been as interesting as currently. The latest generation of VTI accelerometers are used to identify unwanted precedence of human being in vehicle, improving personal safety. Even very small sources of vibration like human heart beat can be detected by VTIs 3D - MEMS sensors with cars body electronic. In further applications the utilization is not limited to detection of human beings hidden in parked cars but a utilization of the same sensor for preventive maintenance diagnosis. This implied that even before fatigue break of mechanical components due to wear-out a car would be in a position for self diagnosis and in combination with telematics a dealer could have spare parts already on stock when a customer enters for regular maintenance leading to more reliable cars and trucks with improved up-time. These applications require resolution and accuracy that has been available only in inclinometers. Through variable resolution of the VTI digital platform this measurement accuracy is available from sensors primarily used by other applications, bringing added value from the sensor fusion to the customer.

In engine vibration compensation the effect of massive engine vibration is significantly reduced by counter - phase vibration fed though accelerometer specifically tuned for engine vibration frequencies. This reduces the efforts for mechanical counter-shafts and makes engines with selective cylinder shutdown more comfortable for the driver.

Active suspension systems are almost as old as suspension systems overall. The systems have traditionally suffered from oversensitivity for small undulations on road surface whereas the mechanical system provided by the steel spring / shock absorber / air filled tire system has been able to filter the small undulations on the road from the driver. The combination of mechanical filtering, achieved through gas damping, together with digital filtering on ASIC level provide means for cost effective, space saving solutions in the area of active or semi-active suspension, enhancing driving properties through reduced un-sprung mass.

2.3 Vibration Suppression

In addition to applications requiring information on smallest dynamic status change of the vehicle dynamics, there are a number of applications suffering from too sensitive measurement. A major challenge for sensor fusion will be the combining of the needs of these two contradicting viewpoints.

Vibration suppression becomes especially critical when mounting position offer high degree of vibration noise and the application asks the same time for highly accurate sensor signals with extremely tight offset – like HSA or EPB functions. As long as the mechanical noise signal is within the dynamic range of the sensor, the required measurement signal can be extracted through means of digital signal processing. If the required signal is orders of magnitude smaller than the environmental noise, like within the ABS-unit where hydraulic valves and other means create high profile of vibration noise. It was derived that a combination of mechanical damping (Fig. 8) of the sensing element mass system with a first order characteristic and digital signal processing filtering is bringing suitable results even in most harsh mounting positions.

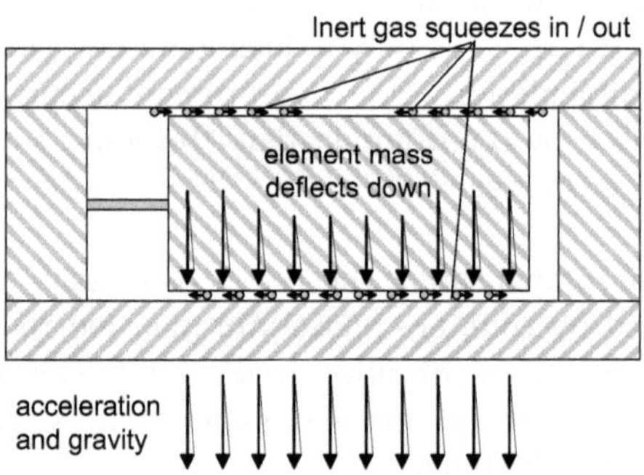

Fig. 8. Vibration suppression by mechanical squeeze film damping in sensing element

Inclinometers in ABS ECUs are one requirement for automotive application. Gear-box inclinometers for trucks are coming with very similar applicational requirements. In these truck applications the ECUs are moving towards the gear-box itself so that beside a harsh vibration profile additionally high temperatures are expected leading to higher sensor requirements. VTI just launched an SCA810 single axis inclinometer which is dedicated for this particular technical challenge. The measurement system performance that can be achieved in this environment is as high as ± 17 mg ($\pm 1°$) in offset stability. In order to guarantee this performance over lifetime comprehensive calibration and test procedure was developed.

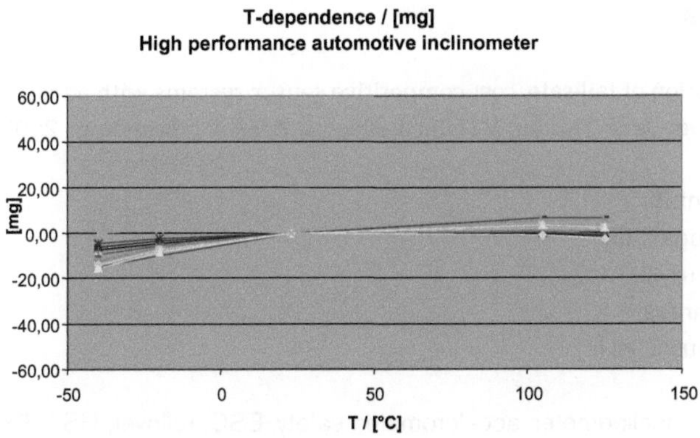

Fig. 9. Offset performance of the VTI digital inclinometer

3 Summary and Outlook

VTI has currently launched the Digital Automotive Platform with Accelerometers and Inclinometers in the first round. This platform is being expended towards multi - axes / multi - *g*-level derivates as well as high accuracy inclinometers. This is achieved by high performance sensing elements in 3D - MEMS technology as well as due to enhanced signal conditioning measurement front-end in combination with excellent digital signal conditioning and filtering and SPI digital output.

With this platform the sensor fusion is becoming reality, both through consumer awareness and cost awareness driven by the legislation in US. ESC sensors, being most performance critical sensors in the vehicle will have secondary responsibilities towards different comfort applications, using the same measurement information for various applications.

References

[1] Realisation of fail-safe, cost competitive sensor systems with advanced 3D - MEMS elements, Jens Thurau, VTI Technologies, AMAA proceedings 2005

Tommi Vilenius
VTI Technologies Oy
Myllynkivenkuja 2
FI-01620 Vantaa
Tommi.Vilenius@ vti.fi

Keywords: inclinometer, accelerometer, safety, ESC, rollover, HSA, EPB, heart beat detection, engine vibration compensation, fail-safe, offset stability

Appendix A

List of Contributors

List of Contributors

Appendix B

List of Keywords

List of Keywords

MIX
Papier aus verantwortungsvollen Quellen
Paper from responsible sources
FSC® C105338

If you have any concerns about our products,
you can contact us on
ProductSafety@springernature.com

In case Publisher is established outside the EU,
the EU authorized representative is:
Springer Nature Customer Service Center GmbH
Europaplatz 3, 69115 Heidelberg, Germany

Printed by Libri Plureos GmbH
in Hamburg, Germany